Graded Exercises and Worked Examples in Physics

to Advanced Level

by

M. NELKON, M.Sc., F.INST.P., A.K.C.
Formerly Head of the Science Department, William Ellis School, London

FIFTH EDITION
(with Multiple Choice Questions)

HEINEMANN EDUCATIONAL BOOKS
LONDON

Heinemann Educational Books Ltd
22 Bedford Square, London WC1B 3HH
LONDON EDINBURGH MELBOURNE AUCKLAND
HONG KONG SINGAPORE KUALA LUMPUR NEW DELHI
IBADAN NAIROBI JOHANNESBURG
EXETER (NH) KINGSTON PORT OF SPAIN

ISBN 0 435 68657 7

© M. Nelkon 1952, 1965, 1968, 1971, 1977
First published 1952
Reprinted ten times
Second Edition 1965
Reprinted 1966
Third Edition 1968
Fourth Edition 1971
Reprinted twice
Fifth Edition 1977
Reprinted 1979, 1980, 1983

Text set in 10/11 IBM Press Roman
Printed and bound in Great Britain by
Richard Clay (The Chaucer Press) Ltd,
Bungay, Suffolk

Preface to Fifth Edition

In this edition the Exercises have been modernized to take account of the latest Advanced level syllabuses of the Examining Boards. The main changes are as follows:

(1) *Mechanics and Properties of Matter.* More updated questions on motion, a separate exercise on simple harmonic motion, and additional questions on intermolecular forces, followed now by the exercise on elasticity.

(2) *Heat.* Electrical heating and measurements are given greater prominence in the questions on heat capacity and latent heat, the mole is now used more in gases, and the exercise on the kinetic theory of gases has been expanded. There is now a reduction in the questions on thermal expansion of solids and liquids.

(3) *Geometrical Optics.* Questions have been modified and the exercise on optical instruments now begins with telescopes.

(4) *Physical (Wave) Optics.* The wave theory questions have been grouped with those on the velocity of light, and there are now separate exercises on interference and diffraction.

(5) *Waves.* Questions cover all matter waves and electromagnetic waves.

(6) *Electrostatics.* There are now more updated questions on electric fields and on capacitors.

(7) *Current Electricity.* More modern questions on magnetic fields and associated phenomena are included, there is a separate exercise on the potentiometer, and the questions on electrolysis are modernized.

(8) *Atomic Physics.* The exercise on photoelectricity now precede that on X-rays and the section concludes with radioactivity and nuclear energy.

Finally, the worked examples at the end of sections have been revised.

I am considerably indebted to M. V. Detheridge, William Ellis School, London, and S. S. Alexander, formerly Woodhouse School, London, for their assistance with this edition, and to Rev. M. D. Phillips, O.S.B., Ampleforth College, York, for his generous assistance with the previous edition on which the new edition is based.

Note to 1983 Reprint

In this reprint, additional modern questions on mechanics, fluid motion, photoelectricity, energy levels and radioactivity have been supplied.

Preface to First Edition

This book contains graded Exercises in Physics. They are intended for the sixth-form student and students at technical colleges taking the General Certificate of Education Advanced level and examinations of similar standard.

The Exercises cover Mechanics, Properties of Matter, Heat, Optics, Waves, Sound, Electrostatics, Current Electricity, Atomic Physics. They are designed not only to provide practice in Physics calculations, but also to test a knowledge of the fundamental points in the descriptive parts of the various subjects. In this way the questions embrace the whole of the subject, and they should be particularly useful for the sixth form or intermediate student on a first study of the topic concerned. The topics may be taken in any order as the teacher desires. Worked examples of Advanced level, or Intermediate standard, have been included at the end of each section to illustrate the fundamental points in the subjects.

Suggestions for the improvement of the book, and notice of errors, would be gratefully received.

Contents

EXERCISES

	Page
MECHANICS	
1. Dynamics. Motion, Momentum, Energy	1
2. Circular Motion. Gravitation	3
3. Simple Harmonic Motion	5
4. Rotational Dynamics	7
5. Statics. Fluids. Miscellaneous Mechanics	9
PROPERTIES OF MATTER	
6. Molecules. Intermolecular Forces	12
7. Elasticity	13
8. Solid Friction. Viscosity	14
9. Surface Tension	15
10. Multiple Choice Questions – Mechanics, Properties of Matter	17
Worked Examples	22
HEAT	
11. Heat Capacity. Latent Heat	26
12. Expansion of Gases. Ideal Gas Equation	28
13. Kinetic Theory of Gases	30
14. Heat Capacities of Gases. Isothermal and Adiabatic Changes	32
15. Vapours. Real Gases	34
16. Thermal Conduction	36
17. Thermal Radiation	37
18. Thermal Expansion of Solids and Liquids	39
19. Thermometry	40
20. Multiple Choice Questions – Heat	41
Worked Examples	46
GEOMETRICAL OPTICS	
21. Reflection at Mirrors	51
22. Refraction at Plane Surfaces	52
23. Refraction through Prisms	53
24. Refraction by Lenses. Defects of Vision	54
25. Dispersion by Prisms and Lenses	56
26. Optical Instruments	58

CONTENTS

PHYSICAL OPTICS
	Page
27. Wave Theory of Light. Velocity of Light	60
28. Interference	62
29. Diffraction	63
30. Polarization	64
31. Multiple Choice Questions – Optics	65
Worked Examples	69

WAVES. SOUND
32. Waves	75
33. Sound	77
34. Waves in Pipes, Strings, Rods	79
35. Multiple Choice Questions – Waves, Sound	81
Worked Examples	82

ELECTRICITY
36. ELECTROSTATICS. Force and Potential	86
37. Capacitors	88
38. Multiple Choice Questions – Electrostatics	91
Worked Examples	93
39. CURRENT ELECTRICITY. Circuit Calculations, Conductors	95
40. Electrical Energy and Power	97
41. Wheatstone Bridge. Resistance	98
42. Potentiometer	99
43. Electrolysis. Cells	101
44. Force on Conductors in Magnetic Fields	102
45. Electromagnetic Induction	104
46. Magnetic Fields of Conductors	106
47. Magnetic Properties of Materials	107
48. A.C. Circuits	108
Worked Examples	110

ATOMIC PHYSICS
49. Electrons and Ions – Motion in Fields	114
50. Diode Valves. Cathode Ray Tube. Junction Diode. Transistor	116
51. Photoelectricity. X-Rays	119
52. Energy Levels in Atoms	121
53. Radioactivity. The Nucleus. Nuclear Energy	122
Worked Examples	125
54. Multiple Choice Questions – Electricity, Atomic Physics	129
55. Multiple Choice Questions – Graphs	135
56. Miscellaneous Multiple Choice Questions	138
ANSWERS	143

1
Mechanics

1. DYNAMICS – MOTION, MOMENTUM, ENERGY

(Assume $g = 10$ m s^{-2} or 10 N kg^{-1})

1 A train moving with a velocity of 72 km h^{-1} (20 m s^{-1}) accelerates uniformly for 20 s at the rate of 1 m s^{-2}. Calculate the distance travelled during this time. Prove any formula used.

2 A ball is thrown vertically upwards with an initial velocity of 10 m s^{-1}. Calculate (i) the maximum height reached, (ii) the velocity at a point half-way up, (iii) the time taken for the ball to return to the thrower.

3 A force of (i) 50 N, (ii) 200 N acts on a mass of 100 kg. What is the acceleration in each case? How long will it take to reach a velocity of 20 m s^{-1} in each case from rest?

4 A ball is thrown at an angle of 15° to the horizontal with an initial velocity of 20 m s^{-1}. Calculate the range of the ball. What is the maximum range possible for a velocity of 20 m s^{-1}?

5 Calculate the kinetic energy and momentum of (i) a bullet of 50 g moving with a velocity of 10 m s^{-1}, (ii) a car of 1000 kg moving with a velocity of 20 m s^{-1}.

6 A force of 50 N acts on an object initially at rest (i) for a time of 4 s, and (ii) for a distance of 20 m. Calculate the momentum and energy gained in the two cases.

7 A ball of mass 0.1 kg is thrown vertically upwards with a velocity of 10 m s^{-1}. Calculate its potential energy and kinetic energy at (i) the maximum height and (ii) the height which is half the maximum.

8 A hose has a hole of cross-sectional area 50 cm^2 and ejects water horizontally at a speed of 0.3 m s^{-1}. If the water is incident on a vertical wall and its horizontal velocity becomes zero, what force is exerted on the wall? At what water speed is the force sixteen times as small?

9 An object of mass 80 kg slides 100 m down an inclined plane at an angle of 30° to the horizontal. Calculate the loss of potential energy of the object. If the coefficient of dynamic friction is 0.2, find the work done against the frictional force, and the kinetic energy acquired by the object.

10 A pile-driver of 200 kg descends from a height of 20 m on to a stake of mass 50 kg. Find (i) the energy of the pile just before it hits the stake, (ii) the common velocity of the stake and pile when they are just moving together, (iii) the loss of energy due to collision in the latter case.

11 Raindrops fall vertically with a velocity of 10 m s^{-1} on to the flat horizontal roof of a shed of area 4 m². Calculate the force on the roof due to the rain, assuming the drops are brought to rest on impact.

What is the force normal to the roof if it slopes at 30° to the horizontal? (Density of water = 1000 kg m^{-3}).

12 A man of mass 60 kg is in a lift which (i) descends, (ii) ascends with an acceleration of 0.3 m s^{-2} in each case. Calculate the reaction of the floor on the man in each case. When is the man 'weightless'?

13 State the *principle of the conservation of momentum*. A mass of 20 kg is suspended at one end of a string, and is struck by an object of 40 kg moving in a downward direction at an angle of 30° to the horizontal with a velocity of 0.1 m s^{-1}. Calculate the velocity of the mass of 20 kg just after the collision if the two weights coalesce.

14 What is an *elastic* and an *inelastic* collision? Does the law of conservation of linear momentum apply to either collision? Explain your answer.

15 An object A of mass m, moving with a velocity of 10 m s^{-1}, collides with a stationary object B of equal mass m, where m is in kg. After the collision, A moves with a velocity u at an angle of 30° to its initial direction and B moves with a velocity v at an angle of 90° to the direction of u.

Applying the conservation of linear momentum in (i) the initial direction of A and (ii) the perpendicular direction, calculate u and v from the two equations obtained.

Calculate the energy of A before collision and the total energy of A and B after collision. Is the collision 'elastic'?

16 An electron of mass 9.0 x 10^{-31} kg is moving with a velocity of 10^6 m s^{-1}. Calculate its kinetic energy. If the electron acquires this velocity from rest in moving through a distance of 0.1 m in a vacuum, calculate the force acting on it.

17 A molecule of gas of mass 10^{-26} kg strikes the plane wall of a cube with a mean velocity normal to the wall of 400 m s^{-1}. Calculate the momentum change at the wall. If there are 12 x 10^{23} molecules in the cube, and one-third are considered to strike this particular wall at a mean interval of 5 x 10^{-4} s, calculate (i) the average force at the wall, (ii) the mean pressure if the area of the wall is 10^{-2} m².

18 A stationary object explodes into two fragments of relative mass 1 : 100. At the instant of break-up, the larger mass has a velocity of 10 m s^{-1}. Calculate (i) the velocity of the smaller mass, (ii) the ratio of their kinetic energies at this instant.

19 10^{12} electrons per second strike a television screen with an average velocity of 10^7 m s^{-1}. If the mass of an electron is 9 x 10^{-31} kg, calculate the force on the screen.

20 What are the dimensions in mass, length and time of: *velocity, acceleration, energy, force, momentum*? Which are vectors?

21 Show that the following relations are dimensionally correct:
(i) $v = \sqrt{T/m}$, where v is the velocity of a transverse wave along a string

whose tension is T and whose mass per unit length is m;

(ii) $v = \sqrt{\gamma p/\rho}$, where v is the velocity of a sound wave in a gas, γ is the ratio of the molar heat capacities, p is the pressure and ρ is the density.

22 State the *principle of the conservation of momentum*. Describe an experiment to verify this principle, showing clearly how your calculations are made.

23 Prove that the total mechanical energy of a ball thrown vertically upwards is constant, neglecting air resistance.

24 What is the relation between *work* and *torque* (*moment*). State the dimensions of work and torque, and explain why they are the same.

25 An object of mass m collides with a stationary object of mass m and both objects move in different directions after the impact. Write down two equations between the masses and the velocities before and after impact if the collision is an elastic one. What is the *angle* between the two directions after impact?

In photographs showing the collision of α-particles with helium nuclei, some forked tracks at 90° to each other are obtained. What do you deduce about the mass of an α-particle?

26 A ball is thrown vertically upwards from the ground. Applying the principle of conservation of momentum to the ball and the Earth, explain what happens to the momentum of the ball and of the Earth as (i) the ball rises to its highest point, (ii) the ball falls to the ground.

2. CIRCULAR MOTION. GRAVITATION.

(Assume, where necessary, that $g = 10$ m s^{-2} or 10 N kg^{-1})

Circular Motion

1 An object of mass 0.05 kg is moving in a circle of radius 0.1 m with a uniform angular velocity of 2 rad s^{-1}. Calculate (i) the speed of the object, (ii) its acceleration towards the centre, (iii) its kinetic energy, (iv) its period of motion.

2 Repeat Question 1 if the object has a mass of 0.01 kg and is moving in a circle of radius 0.4 m at 5 rev s^{-1}.

3 The moon has a period of 27.5 days round the earth. What is its velocity in km per hour, assuming it moves uniformly in a circular orbit of radius 3.8×10^8 m?

4 A small object of mass 0.5 kg attached to the end of a string is whirled round in a horizontal circle of radius 2 m. The string breaks when the tension in it exceeds 100 N. Calculate the maximum angular velocity of the object.

The object is now whirled in a *vertical* circle by the string and its speed is increased from zero. Will the string break when the object is at the top or the bottom of the circle? Give reasons for your answer.

5 Calculate the angle at which a curve must be banked to prevent side-slip at 90 km h^{-1} (25 m s^{-1}) if the radius is 100 m. Prove the formula used.

6 In Question 5, a car of mass 1000 kg travels round the banked track at 25 m s^{-1} without side-slip. Calculate (i) the centripetal force on the car, (ii) the ground reaction on the car.

7 An object at the end of a string 1 m long is whirled round in a horizontal circle so that the string makes an angle of 30° to the vertical. Calculate the speed of the object.

8 An object O of mass 0.2 kg is whirled in a vertical circle at the end of a string of length 1 metre at a steady rate of 5 rev s^{-1}. Find (i) the speed in the circle, (ii) the tension in the string when O is at the top of the circle, (iii) the tension when O is at the bottom of the circle.

9 In a cyclotron, a particle of mass 3 x 10^{-27} kg moves round a circle of radius 0.5 m at 1.2 x 10^7 rev s^{-1}. Calculate (i) the centripetal force, (ii) the kinetic energy of the particle.

10 The hydrogen atom has an electron of mass about 9 x 10^{-31} kg moving in a circular orbit of about 5 x 10^{-11} m. The centripetal force on the electron is about 10^{-7} N. Find (i) the velocity of the electron, (ii) its kinetic energy.

11 An object of mass m moves round a circle of radius r with a constant angular velocity ω. When it moves from A to B, where AB is a diameter of the circle, what change occurs in (i) the speed of the object, (ii) its velocity, (iii) its kinetic energy, (iv) its linear momentum?

With the aid of diagrams, explain why the object has an acceleration towards the centre and prove this is equal to $r\omega^2$.

Gravitation (Assume $G = 6.7 \times 10^{-11}$ N m^2 kg^{-2})

12 Given that the radius of the earth is 6.36 x 10^6 m, at what height above the earth is the value of the gravitational intensity one-quarter its value at the earth's surface assuming the earth is spherical?

13 An earth satellite moves in a circular orbit at a height of 1.0 x 10^6 m above the earth. Calculate the period of motion assuming the earth's radius is 6.4 x 10^6 m.

14 Show that the period T and radius r of the orbit of an earth satellite are related by $T \propto r^{3/2}$.

If the period of a satellite at a distance of 8.0 x 10^6 m from the centre of the earth is 7000 s, calculate the least period of an earth satellite. (Radius of earth = 6.4 x 10^6 m)

15 If the acceleration of free fall, g, is 9.8 m s^{-2} and the mean radius of the earth is 6.4 x 10^6 m, calculate the mass of the earth by considering the gravitational force on an object at the earth's surface.

16 State *Newton's law of gravitation*. Show that $G = gr_E^2/M$ where M is

MECHANICS

the mass of the earth, r_E its radius and g is the acceleration of free fall.

17 The period of rotation of the moon round the earth is 27.3 days. The radius of the earth is 6.4×10^6 m and that of the moon's orbit is 60.1 times as large. Calculate a value for g, and give the theory.

18 Assuming the earth is a sphere of radius 6400 km and $g = 9.8$ m s^{-2}, estimate the mean density of the earth.

19 If the earth is assumed to be a sphere of radius 6.4×10^6 m and density 5500 kg m^{-3}, and Mars to be a sphere of radius 3.4×10^6 cm and density 3900 kg m^{-3}, estimate a value for the acceleration due to gravity on the surface of Mars if that on the earth's surface is 9.8 m s^{-2}.

20 (i) A satellite just skims round the earth in its orbit. Calculate its velocity if the radius of the earth is 6.4×10^6 m and $g = 10$ m s^{-2}. (ii) A satellite in a circular orbit in an equatorial plane is at a height of 36×10^6 m above the earth. Show that the period is about 24 hours. What is the practical significance of this result?

21 The distance between the centre of the earth and the moon is about 400 000 km, and their relative masses are about 81 : 1. If a spacecraft travels away from the earth, at what distance from the centre of the earth does it begin to come mainly under the influence of the moon's attraction?

22 The relative masses of the earth and moon are about 81 : 1 and the earth's radius is 3.7 times that of the moon. Show that the acceleration due to gravity of a falling object on the moon is about $g/6$, where g is the acceleration on the earth.

23 A moon of Jupiter orbits about the planet with a period of 3.5 days in a circle of radius 6.8×10^5 km and the moon orbits about the earth with a period of 27 days in a circle of radius 3.9×10^5 km. What is the ratio of the mass of Jupiter to that of the earth?

24 If the potential due to the earth's gravitational attraction is conventionally taken as 'zero' at an infinite distance away from the earth, the potential at a point outside the earth at a distance r from the centre is given by $-GM/r$, where M is the mass of the earth. (i) Explain the reason for the 'minus', (ii) calculate the speed with which a rocket must be fired from the earth's surface, radius 6400 km, so as to escape the influence of the earth's gravitational attraction. (Hint: use the relation $GM/r^2 = g$.)

3. SIMPLE HARMONIC MOTION

(Assume $g = 10$ m s^{-2} or 10 N kg^{-1})

1 The motion of a mass m is represented by the following expressions, where x is the displacement from a fixed point and k and a are positive constants. Which of these expressions represent simple harmonic motion?
A acceleration = kx, B velocity = kx, C acceleration = $-kx$,

6 GRADED EXERCISES AND WORKED EXAMPLES IN PHYSICS

D velocity $= -k(x^2 - a^2)$, E velocity $= \sqrt{k(a^2 - x^2)}$

2 A mass of 0.01 kg is vibrating up and down with simple harmonic motion; its amplitude is 40 mm, and the period is 2 s. Find (i) the velocity and acceleration at the centre of oscillation, (ii) its velocity and acceleration 10 mm from the centre, (iii) its kinetic energy at the centre. At what point is the *potential energy* at a maximum?

3 In simple harmonic motion, a mass of 0.1 kg has an amplitude of 10 mm. Calculate (i) the maximum velocity and (ii) the maximum force on the mass, if the period of the motion is $\pi/2$ s.

4 A vibrating mass of 0.06 kg has an amplitude of 0.08 m and a period of 4 s. Find the velocity, acceleration and kinetic energy at the middle of the motion. What is the maximum value of the kinetic energy during the motion?

5 Find the length of the simple pendulum which has a period of (i) 2 s, (ii) 0.5 s.

6 Draw sketches of the variations of (i) the kinetic energy, (ii) the potential energy, (iii) the total energy of a particle during one cycle of s.h.m. Account for the appearance of the graphs.

7 The bob of a simple pendulum 1.20 m long weighs 0.500 N. Find the tension in the suspension when the bob passes through the centre of oscillation, if the amplitude is 0.06 m.

8 A helical spring is extended 4 mm by a force of 0.8 N. What is the force constant k ($F = -kx$) of the spring?

The spring is placed on a horizontal smooth surface with one end fixed, and a mass of 0.5 kg is attached to the other end. The mass is pulled 2 mm and released. Calculate (i) the period of its vibration, (ii) the maximum potential energy of the spring, (iii) the maximum kinetic energy of the mass.

9 In a crystalline solid, the vibrating ions are subject to a periodic force whose force constant k is about 60 N m^{-1}. Assuming the mean mass of an ion is 5×10^{-25} kg, estimate its frequency of vibration.

10 A mass of 0.01 kg is suspended from a light spring of original length 0.80 m, and the extension produced is 10 mm. The mass is pulled down 5 mm and then released. Calculate (i) the period of the motion, (ii) the velocity and kinetic energy of the mass as it passes through the centre of oscillation, (iii) the velocity and tension in the spring as the mass moves upward past the point 1 mm from the centre of oscillation. What is the maximum potential energy of the *spring*?

11 A small mass on a vibrating platform has an amplitude of 50 mm. At what frequency does the reaction of the platform become zero at an instant of the cycle and at what point in the cycle does it happen?

12 A U-tube with limbs close together contains oil to a height of 0.2 m in each limb. The density of the oil is 800 kg m^{-3}. After blowing gently down one side the liquid oscillates for a short time. Find the period of oscillation. Derive any formula used. If l is the total length of liquid in the tube, including that in the bend of the U-tube, write down a formula for the period T.

13 A mass m is attached to the end of a light spring on a smooth horizontal table and the other end is fixed to a point on the table. When the mass is displaced a small distance and released, the mass and spring vibrate with S.H.M. The work done in stretching a spring a small distance x is $\frac{1}{2}kx^2$, where k is the 'stiffness factor' of the spring.

(i) If the spring is stretched a distance a when m is at the end of its oscillation, what is the potential energy (p.e.) of the spring and the kinetic energy (k.e.) of the mass at this instant?

(ii) What are the respective values of p.e. and k.e. as the mass passes through the centre of oscillation?

(iii) By energy or other considerations, show that the period of s.h.m. is $T = 2\pi \sqrt{(m/k)}$.

4. ROTATIONAL DYNAMICS

(Assume $g = 10$ m s^{-2} or 10 N kg^{-1})

1 Complete the following table, which shows the analogy between translational (linear) motion and rotational motion:

	Translational motion		Rotational motion	
	Quantity	Formula or Symbol	Quantity	Formula
1.	Mass	m	..1..	I or Σmr^2
2.	Force	$F = ma$	Torque (Couple)	..2..
3.	Linear momentum	mv	..3..	..4..
4.	Work	F x displacement	Work	Torque x ..5..
5.	Kinetic energy	$\frac{1}{2}mv^2$	Kinetic energy	..6..
6.	..7..	F x v	..8..	I x ..9..

2 A disc has a moment of inertia about its centre of 0.1 kg m^2 and rotates about an axis through the centre perpendicular to its plane at 5 rev s^{-1}. Calculate (i) its kinetic energy, (ii) its angular momentum about the centre, (iii) the number of revolutions it makes before coming to rest, and the time taken when a constant opposing couple of 5.0 N m is applied.

3 The electron in a hydrogen atom rotates at a frequency of

6.6×10^{15} rev s^{-1} in a circle of radius 5.3×10^{-11} m. The mass of the electron is 9.1×10^{-31} kg. Calculate the orbital angular momentum of the electron about an axis through the centre of the circle perpendicular to its plane, and its kinetic energy.

4 The earth rotates with a period of 1 year about the sun in an orbit which is roughly circular and of average radius 1.5×10^{11} m. The mass of the earth is 6×10^{24} kg. Calculate the orbital angular momentum of the earth about an axis through the sun perpendicular to the plane of the orbit.

5 A horizontal turntable has a moment of inertia of 0.05 kg m^2 about its centre and rotates steadily at 2.0 rev s^{-1} about an axis through its centre. (i) What is its angular momentum about the centre? (ii) If a small mass of 200 g is placed gently on the rim of the table of radius 50 cm, calculate the reduced frequency of the turntable in rev s^{-1}.

State the Principle used in your calculation.

6 A wheel of moment of inertia 0.04 kg m^2 about its centre rotates steadily at 5 rev s^{-1} about an axis through its centre. When an opposing constant couple of 0.02 N m is applied, the wheel slows down to a final angular rotation of 3 rev s^{-1}. Calculate (i) the time taken for the wheel to reach the angular rotation of 3 rev s^{-1}, (ii) the change in kinetic energy of the wheel, (iii) the number of revolutions made by the wheel during the change from 5 rev s^{-1} to 3 rev s^{-1}.

7 (i) Two masses A and B, on a light horizontal rod, rotate in a horizontal plane in the same direction about the midpoint O of AB with an angular velocity of 10 rad s^{-1} A has a mass of 0.2 kg and B has a mass of 0.1 kg. Calculate their total angular momentum about O if the distance AB is 0.4 m.

(ii) If the mass B slips along the rod so that the distance BO becomes 0.3 m, calculate the new angular velocity of the masses about O.

(iii) If the mass A now slips along the rod so that the distance AO also becomes 0.3 m, calculate the new angular velocity of the masses about O.

8 A wheel of moment of inertia 0.1 kg m^2 is mounted on a horizontal axle of radius 4 cm. A string, partly wound round the axle, has a mass of 0.2 kg attached so that the string hangs vertically. Find (i) the torque on the wheel as the mass is released from rest and falls, (ii) the angular acceleration of the wheel, (iii) the angular velocity of the wheel after 2 s, (iv) the total kinetic energy of the wheel and mass at this instant.

9 Define *moment of inertia*. Derive from first principles the expression for the kinetic energy of a rigid body rotating about an axis with a moment of inertia I and angular velocity ω.

10 A uniform rod 1.2 m long is suspended at one end, inclined at 30° to the vertical, and then released. Calculate the angular velocity with which it passes the vertical. (Moment of inertia of rod of length l about end = mass $\times l^2/3$.)

11 What is the formula for the kinetic energy of (i) a particle, (ii) a rotating body? A hoop of radius 0.6 m rolls from rest without slipping down

MECHANICS 9

an inclined plane 20 m long and angle 30°. Calculate the time taken. (Moment of inertia of hoop about centre = mass x radius².)

12 A hoop of radius 0.6 m oscillates about an axis on its circumference (i) perpendicular, (ii) parallel to the plane of the hoop. Calculate the period in each case.

13 A disc mounted on an axle of radius 2 cm has a moment of inertia of 2×10^{-2} kg m². A force of 1 N is applied tangential to the axle. Find the angular acceleration and the angular velocity after 5 s.

14 A uniform rod 4 m long oscillates through a small angle when suspended from one end. Calculate the period of oscillation. Derive the formula for the period from first principles.

15 A sphere (i) slides down, (ii) rolls down, without slipping, a plane inclined at 30° to the horizontal. Calculate the time taken to travel 10 m along the plane in each case, starting from rest. (Moment of inertia of sphere about axis through centre = $\frac{2}{5}mr^2$.)

16 Describe an experiment to measure the moment of inertia of a flywheel. Give the theory of the method.

5. STATICS. FLUIDS

Statics

1 A picture weighing 40 N is suspended by means of a string passing round a nail, each part of the string making an angle of 60° with the vertical. Calculate the tension in the string.

2 A uniform ladder of weight 200 N and 10 m long rests with one end on the ground at an angle of 60° to the horizontal, with its upper end resting on a smooth wall. Find the reactions at the wall and ground (i) by using a triangle of forces, (ii) by a calculation method.

3 A uniform beam, 6 m long and weight 140 N is hinged at one end to a wall, and is kept horizontal by means of a rope attached to the other end. If the rope is tied to a point 10 m directly above the hinge, calculate the tension in the rope and the reaction at the hinge.

4 An object of 200 N is on the point of slipping on a rough inclined plane which is at 30° to the horizontal. Calculate the frictional force and the coefficient of static friction.

5 A uniform rule 80 cm long is supported at both ends, and weights of 0.6, 0.4 and 0.3 N are placed at distances of 20, 30 and 60 cm from one end. Calculate the reactions at the supports (i) if the weight of the rule is negligible, (ii) if the weight of the rules is 0.5 N.

6 A solid rectangular box of height 100 cm and width 60 cm has a rope attached to the middle of one of its upper sides. The weight of the box is 600 N. If the rope is pulled (i) horizontally, (ii) at an angle of 30° to the horizontal in a downward direction, find the minimum tension in each case required to tilt the box about one edge.

7 Two equal and opposite forces are separated by a distance of 2 m. (i) What name is given to the two forces? (ii) If each force is 80 N, calculate their total moment about a point whose perpendicular distance from one force is x.

What is their total moment about any point if each force is 50 N and they are 1 m apart?

8 A beam balance has three coplanar knife-edges and equal arms 20 cm long. The beam has a mass of 250 g and its centre of gravity is 2 mm below the middle knife-edge. Find the angular deflection of the beam when the masses in the scale-pans differ by 100 milligram.

9 An isosceles triangular plate of height 6 cm and base 4 cm is attached with its base to a square of side 4 cm made of exactly the same metal to make a flat object. Calculate the distance of the centre of mass of the combined object from the apex of the triangle.

10 An open metal cylinder is half-filled with water. The height of the cylinder is 12 cm and its radius is 6 cm. Find the distance from the bottom of the centre of mass of the whole arrangement (i) neglecting the mass of the circular base, (ii) taking the weight of the base into account. (Mass per cm^2 of the cylinder = 5 g.)

11 Discuss the potential energy changes when an object is very slightly displaced when in a position of (i) stable, (ii) unstable equilibrium. Explain the equilibrium of an object pivoted (a) below, (b) above, (c) at its centre of gravity.

Fluids (Assume 1 N = weight of 100 g mass)

12 The density of aluminium is 2300 kg m^{-3}, the density of copper is 6500 kg m^{-3}, and the density of a mixture of the two metals is 4500 kg m^{-3}. Calculate (i) the ratio of the volumes, (ii) the ratio of the masses used to make the mixture, assuming no change in volume.

13 A brass solid of 100 g mass is totally immersed in (i) water, (ii) oil of density 800 kg m^{-3}. Find the upthrust on the solid in each case if the density of brass is 8500 kg m^{-3}.

14 A piece of aluminium, density 2300 kg m^{-3}, is totally immersed in (i) water, (ii) brine of density 1100 kg m^{-3}. If the aluminium has a volume of 100 cm^3 and is suspended by a thread in the liquid, calculate the tension in the thread in each case.

15 Calculate the density of a copper sulphate crystal from the following measurements:
Weight of crystal - 0.052 N,
weight of crystal in oil of density 800 kg m^{-3} = 0.032 N.

MECHANICS

Fluid Motion (Assume that $g = 10$ m s^{-2}, density of water $= 1000$ kg m^{-3} and the Bernoulli equation applies.)

16 Water, with negligible viscosity, flows through a horizontal pipe of varying cross-section with a steady flow. At a cross-section X of 8 cm^2 the velocity is 3 cm s^{-1}. What is (i) the velocity at a place Y of cross-sectional area 2 cm^2? (ii) the pressure difference between X and Y?

17 In a uniform horizontal pipe of cross-section area 30 cm^2 the static pressure in water flowing steadily is 6.5×10^4 Pa and the total pressure is 6.7×10^4 Pa. Calculate the flow velocity of the water and the mass rate of flow across a section of the pipe.

18 Water flows steadily along a horizontal pipe which has a cross-section of area 20 cm^2 at one place and 5 cm^2 at a narrower place where the water speed is 4 m s^{-1}. The pressure in the narrow part is 4.80×10^4 Pa. Calculate the pressure in the wider part.

19 An open tank contains water 0.8 m deep. A hole of cross-section area 5 cm^2 is made at the bottom of the tank. Estimate the mass of water per second flowing initially out of the hole.

MISCELLANEOUS QUESTIONS

20 A shell of mass 3M, moving in a horizontal direction OX with a velocity of 200 m s^{-1}, explodes into a mass 2M moving with a velocity of 300 m s^{-1} at 60° to OX and a mass M moving with velocity v at an angle θ to OX. Calculate v and θ.

21 (i) A constant torque of 200 N m turns a wheel of moment of inertia 40 kg m^2 about its centre. Calculate the kinetic energy gained in 4 s. If the torque is removed and a constant opposing torque of 50 N m immediately applied, how many revolutions does the wheel make before coming to rest?

(ii) A hollow cylinder, mass 2 kg, radius 0.5 m and moment of inertia 0.5 kg m^2 about its axis, rolls down a plane inclined at 30° to the horizontal without slipping. Calculate the linear acceleration of the cylinder down the plane.

22 A satellite S is launched from the earth E directly towards the moon M. Draw sketch graphs showing roughly how (i) the resultant gravitational force F on S and (ii) the resultant gravitational potential V of S, varies with the position of S between E and M. (Remember that $F = -dV/dr$ and that V is negative between E and M.)

23 One end of a horizontal spring S of force constant 10 N m^{-1} is attached to a fixed point on a smooth table and a mass M of 0.2 kg is attached to the other end of S on the table. M is now displaced 0.04 m away from the spring and released.
Show that the motion of M is simple harmonic and calculate (i) the maximum energy of the spring, (ii) the maximum velocity of M, (iii) the maximum acceleration of M.

2
Properties of Matter

6. MOLECULES. INTERMOLECULAR FORCES

1 The density of copper is about 8500 kg m^{-3} and the mass of 1 mole is 63.6 g. Estimate the number of 'free electrons' per metre^3 if copper has one 'free' electron per atom. (Assume Avogadro constant = 6 × 10^{23} mol^{-1}).

2 Assuming the density of gold is 2 × 10^4 kg m^{-3} at room temperature and the molar mass is 0.2 kg, estimate the mean separation of the molecules. (Avogadro constant = 6 × 10^{23} mol^{-1}).

3 The mass of 1 mole of water is 18 g and its density is 1000 kg m^{-3}. Estimate the mean separation of the molecules in water. (Avogadro constant = 6 × 10^{23} mol^{-1}).

4 Fig. 6A shows roughly the variation of the intermolecular potential V of two molecules with a separation r in an assembly of molecules, and the variation of the intermolecular force F.

(i) From $F = -dV/dr$, why has F positive and negative values?

(ii) If negative values of F represent an attractive force, what do you know about F as r is decreased from OB to OA?

(iii) At what value of r do the attractive and repulsive forces balance?

(iv) From an energy viewpoint, why is the value of r in (iii) considered as the equilibrium separation of the molecules?

(v) From Fig. 6A, what value of r may correspond to the gaseous phase of the molecules?

5 In Fig. 6A, what features of the $F-r$ curve indicate respectively (i) the reason for Hooke's law in elasticity of metals, (ii) a change of state, (iii) thermal expansion of a solid? Explain your answers.

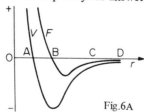

Fig.6A

6 In sodium chloride, the ions in the crystalline state have an interatomic potential V given roughly by

$$V = \frac{a}{r^9} - \frac{b}{r}$$

where r is the separation of the ions and a and b are constants.
Using this relation to find F, the interatomic force, determine the *equilibrium separation* of the ions.

7 Most metals expand on heating but some alloys contract on heating. Draw sketches showing the V–r curves of the two types of solids and explain their difference in shape.

8 When an additional force ΔF is applied to a wire of cross-sectional area A and length l, the length increases by Δl. The tensile stress $\Delta F/A$ has then produced a tensile strain $\Delta l/l$.

From the variation of F with r, the interatomic separation, in Fig. 6A, show that the Young modulus (defined as 'tensile stress/tensile strain') for a metal solid such as copper *decreases* with temperature rise.

7. ELASTICITY

(Assume $g = 10$ N kg^{-1})

1 (i) State *Hooke's law*, (ii) define *tensile stress, tensile strain, Young modulus*; state the units of each quantity.

2 A load of 20 N on a wire of diameter 1 mm produces an extension of 0.1 mm. Calculate the Young modulus for the material of the wire if its length is 3 m.

3 Find the Young modulus for a wire of length 2 m and diameter 2.0 mm if a load of 10 N produces an extension of 0.36 mm.

4 What force will produce an extension of 0.3 mm in a steel wire 4 m long and 2 mm diameter? (Young modulus of steel = 2×10^{11} N m^{-2}.) Calculate the energy stored in the stretched wire.

5 A load of 15 N produces an extension of 0.1 mm in a wire 10 m long. If the Young modulus is 1.8×10^{11} N m^{-2}, find the diameter of the wire.

6 The metal of a wire rotating about one end has a breaking stress of 8×10^9 N m^{-2}. Calculate the maximum number of revs per second of the wire if its mass is 0.4 kg, its length is 2.0 m and its cross-sectional area is 8×10^{-6} m^2.

7 A steel rod of cross-section 0.2 cm^2 is fixed at both ends. The rod is heated from 12°C to 100°C. Calculate the force required to prevent the rod expanding. (Young modulus of steel = 2×10^{11} N m^{-2}; linear expansivity = 1.1×10^{-5} K^{-1}.)

8 Define the terms *proportional limit, elastic limit, elasticity, yield point*. Draw a graph showing how the extension of a wire varies with the applied force, and mark the elastic limit and yield point on it. Explain how the magnitude of the Young modulus is obtained from the graph.

9 Draw a labelled diagram of the apparatus to measure the Young modulus for a wire, and describe the experiment. Show how the Young

modulus is calculated from the readings.

10 A mass of 0.4 kg is placed on the mid-point of a horizontal wire 2 m long fixed at each end, and the mid-point is depressed 20 mm. Find Young modulus for the wire if its diameter is 2 mm.

11 A brass wire of circular cross-section is heated to 200°C and is then prevented from contracting when its temperature falls to 10°C. Find the force required if the diameter of the wire is 0.2 mm. (Young modulus for brass = 1.7×10^{11} N m^{-2}; linear expansivity of brass = 1.8×10^{-5} K^{-1}.

12 Calculate the energy per m^3 of a wire when a load of 5 N is attached to it, if the extension is 4 mm. The original length of the wire is 1 m and its diameter is 6 mm.

13 Calculate the depression of the mid-point of a horizontal steel wire 2 m long fixed at each end, when a mass of 1 kg is placed on it. (Young modulus = 2×10^{11} N m^{-2}; diameter of wire = 1 mm).

14 Define *bulk modulus, shear modulus*. With diagrams, explain the difference between bulk and shear stresses and strains. Show that, for bulk modulus, the strain energy per unit volume is ½ x stress x strain for a material which obeys Hooke's law.

8. SOLID FRICTION. VISCOSITY

(Assume $g = 10$ N kg^{-1})

Solid Friction

1 Define *coefficient of static friction* and *coefficient of dynamic (kinetic) friction*. Which has the greater magnitude? An object of mass 0.5 kg rests on a horizontal surface, and a force of 2.2 N is required to start it moving with constant velocity. What is the coefficient of dynamic friction?

2 A uniform ladder 10 m long rests with the upper end against a smooth wall, and the lower end rests on the ground at 60° to the horizontal. If the ladder is on the point of slipping, calculate the coefficient of static friction.

3 State the *laws of solid friction*. Describe an experiment to investigate the law relating to the normal reaction and the area in contact with the surface.

4 An object of mass 10.0 kg rests on a rough horizontal plane of coefficient of static friction 0.4. Calculate the *least* force required to move the object along the plane, assuming that the least force is obtained when it acts at an angle θ to the horizontal given by tan θ = 0.4.

Viscosity

5 Define *velocity gradient, coefficient of viscosity*. What are their *units*? Calculate the retarding force on the curved surface of a cylinder of water of radius 5 cm and length 10 cm if the whole of the surface is in a region of velocity gradient 5 s^{-1}. (Viscosity of water = 1.1×10^{-3} N s m^{-2}.)

PROPERTIES OF MATTER

6 How do the laws of fluid friction differ from those of solid friction? Derive the dimensions of coefficient of viscosity in mass, length, time, and name a unit of viscosity.

7 The retarding force F of a sphere moving through a liquid depends on the radius a of the sphere, the coefficient of viscosity η of the liquid, and the velocity v of the sphere. Using the method of dimensions, prove that $F = ka\eta v$, where k is a number.

8 A small steel ball-bearing A falls with terminal velocity through a viscous liquid X. Another steel ball-bearing falls with terminal velocity through a viscous liquid Y.

Calculate the ratio of the viscosities of X and Y if the radii of A and B are in the ratio 2 : 3 and their respective times of fall through the same distance are in the ratio 4 : 3.

9 A small ball-bearing of 2 mm diameter is dropped gently into the middle of a viscous liquid. If the terminal velocity is 4 mm s^{-1}, the density of the bearing is 8000 kg m^{-3} and that of the liquid is 2000 kg m^{-3}, calculate the viscosity of the liquid.

10 A small steel ball-bearing is dropped gently into the middle of a viscous liquid. Draw a sketch showing roughly how its velocity varies with time, and indicate on it the terminal velocity.

11 In a Millikan oil-drop experiment, the drop falls with a terminal velocity of 0.12 mm s^{-1} through air of viscosity 1.8×10^{-5} N s m^{-2}. Calculate the radius of the drop if the density of oil = 900 kg m^{-3} and the density of air is negligible.

12 What is *laminar flow, turbulent flow*? How may laminar and turbulent flow be demonstrated? Using the method of dimensions, show that, under steady flow, the volume of liquid per second flowing along a pipe of radius a under a uniform pressure gradient g is kga^4/η, where η is the coefficient of viscosity.

13 In an experiment to measure the viscosity of water by flow through a horizontal tube, the excess pressure at the ends of the tube was equivalent to a head of 30 cm of water, the diameter and length of the tube were respectively 0.4 mm and 30 cm, and the volume of water issuing per second was 6×10^{-3} cm^3. Using 'volume per second = $\pi p a^4/8\eta l$, calculate a value for the viscosity of water.

14 On the molecular theory, the viscosity of a *gas* is considered due to momentum changes. Explain the theory, and draw a diagram to illustrate your answer.

9. SURFACE TENSION

(Assume $g = 10$ m s^{-2})

1 What are the *units* and *dimensions* of surface tension? How does a liquid surface tension vary with increasing temperature?

2 Define *surface tension*. A rectangular glass slide, 4.0 cm by 1.5 cm, rests in a horizontal position touching a water surface. Find the downward force on it due to surface tension if this is 7×10^{-2} N m^{-1}.

3 A three-sided rectangular frame ABCD is suspended vertically by BC from one arm of a balance, and the counterpoise is 0.58 g. When a soap film is formed in the frame with BC horizontally by lowering the 'open' side AD into soap solution, the new counterpoise is 0.81 g. Calculate the surface tension of the film if BC is 4.6 cm and AB is 3.0 cm long.

4 Water in a capillary tube rises 4.67 cm above the outside level. If the diameter of the tube is 0.64 mm calculate the surface tension of the water, giving the theory. What is the diameter of a capillary tube in which the water rises 8.49 cm above the outside level?

5 A mercury thread in a uniform capillary tube is 12.2 cm long and has a mass 0.242 g. Calculate the height to which the water would rise above the outside level, if the tube were dipped into water. (Assume surface tension of water = 7.2×10^{-2} N m^{-1}, density of mercury = 13 600 kg m^{-3}.)

6 Two capillary tubes, A, B, are dipped into a liquid, which then rises 7.6 and 4.4 cm respectively above the outside level. Compare the diameters of the tubes. Give the theory.

7 How high does an oil of density 800 kg m^{-3} rise in a capillary tube of diameter 0.2 mm if its surface tension is 6×10^{-2} N m^{-1} and the angle of contact is 30°? Give the theory.

8 Calculate the depth to which mercury is depressed inside a capillary tube of 1.0 mm diameter. (Surface tension of mercury (0.54 N m^{-1}, angle of contact with glass = 132°.) Prove the formula used.

9 A soap bubble has a diameter of (i) 4 cm, (ii) 5 mm. Calculate the excess pressure inside the bubble over that outside in each case, giving the theory. (Assume surface tension of soap film = 2.5×10^{-2} N m^{-1}.)

10 A U-tube consists of a tube of 4 mm radius connected to one of 2 mm radius. Calculate the difference in levels of water inside the U-tube if the tension of water is 7×10^{-2} N m^{-1}. Give the theory.

11 Two tubes, A, B, connected by a T-piece, are placed respectively in water and in oil of density 800 kg m^{-3}. If the water is drawn up 8.6 cm in A, how high does it rise in B? (Surface tension of water and oil = 7×10^{-2} and 10^{-1} N m^{-1}; assume the angle of contact zero in each case. Diameter of A = 5 mm = 2 × diameter of B.) Give the theory.

12 Describe an experiment to measure the surface tension of water. Give the theory.

13 Derive the formula for the excess pressure in a bubble in a liquid over that outside.

A small and a large soap-bubble are blown respectively at the ends of two tubes, which are then connected at their other ends so that air can flow between the bubbles. Describe and explain what happens to each bubble.

14 In a standard (Jaeger) method of measuring the variation of surface

PROPERTIES OF MATTER

tension with temperature, a bubble is blown slowly on the end of a capillary tube below the surface of the liquid which is contained in a beaker.
(i) Why is this method superior to that of the 'rise in capillary tube' method?
(ii) Write down the formula for calculating the surface tension.
(iii) As the bubble is blown at the end of the capillary tube, its radius increases until it becomes equal to that of the interior radius of the capillary tube and then breaks away. Explain why the bubble cannot grow further.

What are the missing words in the following questions 15–17:

15 Molecules in the surface of a liquid have . . . potential energy than those in the bulk of the liquid.

16 The potential energy of a liquid surface is a . . . for a given volume of liquid; hence the shape of a drop under surface tension forces is a . . .

17 The 'surface energy' is the isothermal work done in breaking bonds with surrounding molecules in the surface, when all the surface molecules are removed to the vapour state. (i) The surface energy is thus similar to the latent heat of . . . of the liquid. (ii) The surface tension vanishes at the . . . temperature of the liquid.

18 The surface tension γ of a liquid may be defined as 'the energy to increase the surface area by unit amount under isothermal conditions' or 'the free surface energy' of the liquid.

From this definition, (i) derive an expression for the surface energy of a *soap-bubble* of radius r and surface tension γ (assume it grows from zero radius), (ii) find the radius of the soap-bubble formed by joining two soap-bubbles of radii 30 mm and 40 mm respectively under isothermal conditions.

19 A small drop of mercury has a radius of 2 mm. It is divided isothermally into two drops of equal volume. Calculate the mechanical energy required assuming this is equal to the change in surface energy. (Assume γ for mercury = 0.5 N m^{-1}).

10. MULTIPLE CHOICE QUESTIONS – MECHANICS, PROPERTIES OF MATTER

For each of the questions 1–16, choose one statement from **A** *to* **E** *which is the most appropriate.*

1 The period of oscillation of a body suspended by a string of length l is $2\pi\sqrt{l/g}$. This expression is accurate only for
 A a light body and no friction,
 B small amplitude of oscillation and a small body,
 C a body of low density,
 D a body which is not ferromagnetic and large amplitude of oscillation,

E a pendulum at the north or south pole.
2 An object of mass 2 kg is brought to rest from a velocity of 10 m s^{-1} in 5 seconds by a constant resistive force F. In this case
  A $F = 2$ N,
  B the initial energy of the object is 20 J,
  C the initial momentum is 20 J,
  D $F \times 5 = \frac{1}{2}.2.10^2$,
  E $F \times 5 = 2 \times 10$.
3 An object of mass 4 kg and moving at 5 m s^{-1} collides with a stationary object of mass 6 kg and both move off with the same velocity.
  A The common velocity is 2.5 m s^{-1} and momentum is conserved,
  B the common velocity is 2 m s^{-1} and energy is conserved,
  C the common velocity is 2 m s^{-1} and the momentum is conserved,
  D the common velocity v is calculated from $6 \times v = 4 \times 5$,
  E the common velocity is calculated from the conservation of energy principle.
4 A mass M is attached to one end of a light spring S on a smooth table, and the other end of S is fixed. When M is pulled slightly and then released. M and S undergo simple harmonic oscillation. In this case
  A M and S oscillate out of phase,
  B M has maximum kinetic energy when S has maximum potential energy,
  C M has maximum kinetic energy equal to twice the maximum potential energy of S,
  D the total energy of M and S is always constant,
  E M has minimum kinetic energy when S has its normal length.
5 A dancer on ice spins faster about the same axis when she folds her arms. This is due to
  A increase in energy and increase in angular momentum,
  B increase in energy and decrease in angular momentum,
  C decrease of friction at the skate,
  D constant angular momentum and increased kinetic energy,
  E constant angular momentum and constant kinetic energy.
6 A wheel has a moment of inertia about a fixed axis of 0.2 kg m^2, and is rotating with a constant angular velocity of 2 rad s^{-1} about the axis. In this case
  A the angular momentum about the axis is 0.8 kg m^2 rad s^{-1},
  B the kinetic energy is 0.8 J,
  C the wheel can be brought to rest in 0.5 s by an opposing constant torque of 4 N m,
  D the wheel can be brought to rest in a time t calculated from $0.2 \times 2 = 4t$ when the opposing torque is 4 N m,
  E the radius of the wheel is 0.2 m.
7 Which of the following are true:
  A Work is a scalar and pressure is a scalar,
  B Work is a scalar and pressure is a vector,
  C Energy = work done = force × distance, and hence energy is a vector,
  D Linear momentum and density are both vectors

PROPERTIES OF MATTER

E Linear momentum is a scalar.

8 The sensation of 'weightlessness' in a spacecraft in orbit is due to
  A the absence of gravity,
  B the absence of linear motion,
  C the acceleration in orbit equal to the acceleration of gravity outside,
  D the presence of gravity outside but not inside the spacecraft,
  E the spacecraft in orbit has no energy.

9 A lift is accelerating upwards at 2 m s^{-2} and a man inside has a mass of 60 kg. Assuming $g = 10 \text{ N kg}^{-1}$, the reaction on the man at the floor is

    A 120 N   B 600 N   C 700 N   D 720 N   E 1200 N

10 In connection with intermolecular forces,
  A a rubber eraser is difficult to compress owing to the attractive forces between molecules,
  B an extended steel wire recovers in length owing to the repulsive forces between molecules,
  C the potential energy of its molecules is decreased when a wire is stretched,
  D when a solid is in equilibrium the intermolecular distance is that when attractive and repulsive forces balance,
  E there are only attractive forces between molecules.

11 When a light thin ring is placed gently on a clean water surface so that it floats,
  A the surface tension force on the ring is horizontal,
  B the surface tension force on the ring is $\pi r^2 . \gamma$, where r is the radius and γ is the surface tension,
  C the surface tension force on the ring is $2.2\pi r . \gamma$,
  D the surface tension force on the ring acts vertically downwards,
  E the weight of the ring $mg = \pi r^2 . 2\gamma$, since there are two sides to the ring.

12 In the capillary tube method of measuring the surface tension of water,
  A the radius r is best determined by cutting the tube at the meniscus and measuring it with a travelling microscope,
  B r can be found by measuring the mass m and length l of a mercury thread and using the relation $r = \sqrt{m/2\pi l \rho}$, where ρ is the mercury density,
  C the radius r is related to the height h of liquid drawn up in the tube by $r = 2\gamma/hg$,
  D the microscope is first focused on the outside level of water in contact with the tube at the bottom,
  E a concave spherical meniscus is seen through the microscope since the angle of contact is zero.

13 When a long wire, suspended from a fixed point, is extended by a load less than the elastic limit, then
  A the tensile stress varies along the wire,
  B the extension at each section is proportional to its distance from the point of suspension,

20 GRADED EXERCISES AND WORKED EXAMPLES IN PHYSICS

C Young modulus is equal to the ratio tensile strain/tensile stress,
D the yield point may be reached,
E the tensile stress depend on the length of wire.

14 A load of 40 N, which is less than the elastic limit, is attached to a wire 2 metres long and cross-sectional area 10^{-6} m^2, and Young modulus = 2×10^{11} N m^{-2}.
  A The extension is 10^{-3} m,
  B the energy in the wire is 2×10^5 J,
  C the stress is 4×10^5 N m^{-2},
  D the strain is 10^{-4} per centimetre,
  E the energy in the wire is 8×10^{-3} J.

15 When a sphere of radius a and density ρ is dropped gently into a medium of infinite extent and viscosity η and density σ,
  A the terminal velocity $v_0 = 2(\rho - \sigma)ga^2/\eta$,
  B the terminal velocity is calculated from $6\pi\eta a v_0^2 = 4\pi a^3 \rho g$,
  C the velocity of the sphere decreases as it falls owing to friction,
  D the velocity increases to a value $v_0^2/2$,
  E an oil drop accelerates continuously as it falls through air.

16 The same liquid flows uniformly through two capillary tubes, X, Y, of equal length whose respective diameters are in the ratio 2:1. If the respective pressure differences between the ends of the tubes are in the ratio 1:2, the ratio of the volume per second of liquid flowing in X to that in Y is
  A 1 : 1, B 4 : 1, C 8 : 1, D 16 : 1, E 2 : 1.

17 The moment of inertia of the uniform rod LM about the centre is 16 kg m^2 and the force X = 50 N. Forces in the direction of X are counted as positive. Calculate the forces Y and Z in the following cases (a), (b), (c). Choose the value of Y from the five alternatives numbered (i) and the value of Z from those numbered (ii)

Fig. 10A.

(a) Rod is stationary.
  (i) A 50 N B −50 N C 100 N D −100 N E None of these
  (ii) A 50 N B −50 N C 100 N D −100 N E None of these
(b) Rod moves with a uniform linear acceleration of 2 m s^{-2}.
  (i) A 296 N B −296 N C −4 N D −4 N E None of these
  (ii) A 100 N B −100 N C 200 N D −200 N E None of these
(c) Rod has an initial angular acceleration of 3 rad s^{-2} about the centre.
  (i) A 52 N B −48 N C 50 N D 48 N E −52 N
  (ii) A 48 N B −2 N C −2 N D −1 N E None of these

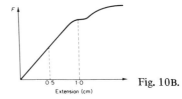

Fig. 10B.

PROPERTIES OF MATTER

18 The graph in Fig. 10B is that of force against extension for metal wire. Which of the following statements is true?
- A Energy to stretch 0.5 cm is 50 J and that to stretch 1.0 cm is greater than 100 J.
- B Energy to stretch 0.5 cm is 50 J and that to stretch 1.0 cm is less than 100 J.
- C Energy to stretch 0.5 cm is 25 J and that to stretch 1.0 cm is greater than 50 J.
- D Energy to stretch 0.5 cm is 25 J and that to stretch 1.0 cm is less than 50 J.
- E None of these.

19 An object of mass 10 kg rests on a pair of bathroom scales in a lift. The lift accelerates upwards with an acceleration of 2 m s^{-2}. Assuming g = 10 m s^{-2}, what are the values of (a) the weight of the object, (b) the resultant force on the object, (c) the reaction of the scales on the weight, (d) the force on the scales causing them to read. Count upward forces as negative and choose answers from

A +20 N   B +100 N   C −120 N   D −20 N   E +120 N

20 A spherical ball bearing falls from the surface of a viscous liquid. Which of the graphs below best describes the variation with time (horizontal axis) of the following quantities, relating to the sphere: (a) Viscous drag, (b) acceleration, (c) apparent weight, (d) resultant force, (e) velocity, (f) distance fallen from the surface.

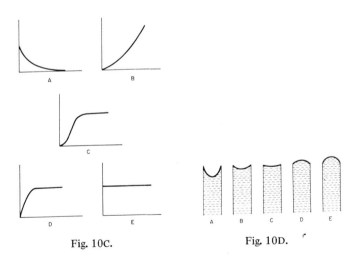

Fig. 10C.   Fig. 10D.

21 The capillary rise of water in a certain tube is 5 cm. If only 4 cm of the tube is above the surface of the water what will the meniscus at the top of the capillary tube look like? Choose your answer from A, B, C, D, E in Fig. 10D.

WORKED EXAMPLES ON MECHANICS AND PROPERTIES OF MATTER

1 A mass of 0.2 kg revolves in a horizontal circle at the end of a string 1 m long, the other end being fixed. The breaking tension of the string is 15 N. Find the maximum speed of the mass.

The acceleration towards the centre = v^2/r, where v is the maximum speed and r is the radius. The force required is thus mv^2/r.

$$\therefore \frac{0.2v^2}{1} = 15$$

$$\therefore v = \sqrt{15/0.2} = 8.7 \text{ m s}^{-1}$$

2 Assuming that the moon revolves uniformly in a circle round the earth's centre, find the acceleration due to gravity at the earth's surface given the radius of the earth = 6.4×10^6 m, the radius of the moon's orbit = 3.84×10^8 m, and the period of rotation of the moon = 27.3 days.

At the earth's surface, the weight of a mass m is equal to the force of attraction of the earth on it.

$$\therefore mg = \frac{GmM}{r_E^2},$$

where G is the gravitational constant, M is the mass of the earth, and r_E is the radius of the earth.

$$\therefore g = \frac{GM}{r_E^2} \qquad \qquad \ldots \ldots \ldots \text{(i)}$$

Suppose m' is the mass of the moon, and r is the radius of the moon's orbit. Then, the force towards the centre on the moon = $m'v^2/r = m'r\omega^2$, where ω is the angular velocity. But the force on the moon due to the earth = $Gm'M/r^2$.

$$\therefore m'r\omega^2 = \frac{Gm'M}{r^2}$$

$$\therefore r\omega^2 = \frac{GM}{r^2} \qquad \qquad \ldots \ldots \ldots \text{(ii)}$$

Dividing (i) by (ii)

$$\therefore \frac{r\omega^2}{g} = \frac{r_E^2}{r^2}$$

$$\therefore g = \frac{r^3 \omega^2}{r_E^2}$$

But $\qquad \omega = \frac{2\pi}{\text{period}} = \frac{2\pi}{27.3 \times 24 \times 3600}$

$$\therefore g = \frac{(3.84 \times 10^8)^3 \times 4\pi^2}{(27.3 \times 24 \times 3600)^2 \times (6.4 \times 10^6)^2} = 9.8 \text{ m s}^{-2}$$

3 An object of mass 0.2 kg vibrates in simple harmonic motion with an amplitude of 15 mm. Its maximum velocity is 0.06 m s^{-1}. Calculate (i) the period of oscillation, (ii) the maximum force on the object.

If r is the amplitude, maximum velocity = $r\omega$. So

$$15 \times 10^{-3}\, \omega = 0.06$$

$$\therefore \omega = \frac{0.06}{15 \times 10^{-3}} = 4 \text{ rad s}^{-1}$$

(i) period $T = \dfrac{2\pi}{\omega} = \dfrac{2\pi}{4} = 1.6$ s

(ii) Maximum acceleration $a = \omega^2 r = 4^2 \times 15 \times 10^{-3}$

$$= 0.24 \text{ m s}^{-2}$$

maximum force = ma = 0.2 × 0.24 = 0.048 N

4 A wheel rotates about an axis through its centre perpendicular to its plane with a constant angular velocity of 10 rad s^{-1}. The moment of inertia of the wheel about the axis is 0.1 kg m^2. (i) Calculate the kinetic energy of the wheel, (ii) if a constant opposing couple of 0.02 N m is exerted against the wheel, find the number of revolutions made before the wheel comes to rest and the time taken.

(i) Kinetic energy = $\tfrac{1}{2} I\omega^2 = \tfrac{1}{2} \times 0.1 \times 10^2 = 5$ J . . . . . . (i)

(ii) Work done by opposing couple = couple × angle of rotation, θ

$$\therefore 0.02 \times \theta = \text{loss in K.E. of wheel} = 5$$

$$\therefore \theta = 250 \text{ rad}$$

$$\therefore \text{no. of revs} = \frac{250}{2\pi} = 40 \text{ approx.} \quad \ldots \quad \text{(ii)}$$

Angular retardation = $\dfrac{\text{couple}}{\text{moment of inertia}} = \dfrac{0.02}{0.1} = 0.2$ rad s^{-2}

Since final angular velocity of wheel = 0,

$$\therefore 0 = 10 - 0.2t$$

$$\therefore t = 50 \text{ s} \quad \ldots \ldots \quad \text{(iii)}$$

(Alternatively, opposing couple × time = change in angular momentum

$$\therefore 0.02 \times t = 0.1 \times 10 = 1$$

$$\therefore t = 50 \text{ s}$$

5 A metal block of weight 20 N and volume 8 × 10^{-4} m^3, completely immersed in oil of density 700 kg m^{-3}, is attached to one end of a vertical wire of length 4.00 m whose other end is fixed. The length of the wire then increases by 1 mm. If the diameter of the wire is 0.6 mm, calculate (i) its Young modulus, (ii) the energy stored in the wire. (Assume g = 10 N kg^{-1}.)

Upthrust on metal = weight of oil displaced = $8 \times 10^{-4} \times 700 \times 10$
= 5.6 N.
Hence force on wire, F, = $20 - 5.6 = 14.4$ N.
Also $l = 4$ m, $e = 1 \times 10^{-3}$ m, $A = \pi r^2 = \pi \times 0.3^2 \times 10^{-6}$ m^2

(i) $\therefore E = \dfrac{F/A}{e/l} = \dfrac{F \cdot l}{A \cdot e}$

$= \dfrac{14.4 \times 4}{\pi \times 0.3^2 \times 10^{-6} \times 1 \times 10^{-3}}$

$= 2.0 \times 10^{11}$ N m^{-2}

(ii) Energy stored = $\tfrac{1}{2} Fe$

$= \tfrac{1}{2} \times 14.4 \times 1 \times 10^{-3}$

$= 7.2 \times 10^{-3}$ J

6 A steel ball, radius 1 mm and density 8500 kg m^{-3}, takes 4.5 s to fall a distance 0.10 m while moving with terminal velocity through a column of liquid X, density 1500 kg m^{-3}. Assuming Stokes' law, $F = 6\pi\eta av$, calculate the coefficient of viscosity of the liquid.

How long will it take a steel ball of radius 3 mm to fall the same distance with terminal velocity through a liquid Y whose coefficient of viscosity is twice that of X? ($g = 10$ m s^{-2}).

At terminal velocity v_0,

weight of ball − upthrust = upward frictional force

$$\tfrac{4}{3}\pi a^3 (\rho - \sigma) g = 6\pi\eta a v$$

$$\therefore \eta = \frac{2(\rho - \sigma) g a^2}{9 \ v_0} \qquad (1)$$

$$= \frac{2(8500 - 1500) \times 10 \times (1 \times 10^{-3})^2}{9 \times 0.1/4.5}$$

$= 0.7$ N s m^{-2}

From (1), the terminal velocity $\propto a^2/\eta$. So terminal velocity-v_0' of ball of radius 3 mm is given by

$$\frac{v_0'}{v_0} = \frac{3^2/1^2}{2/1} = \frac{9}{2}$$

$\therefore$ time to fall same distance $= \dfrac{2}{9} \times 4.5$ s $= 1.0$ s

7 Derive a formula for the excess pressure inside a bubble in a liquid over that outside the bubble. A vertical tube of radius 1 mm is connected by tubing to form a U-tube with a vertical tube of radius 0.6 mm. Oil of density 800 kg m^{-3} is then poured into the U-tube. Calculate the difference in levels in the two tubes if the surface tension of the oil is 0.05 N m^{-1} and the angle of contact is 20°.

PROPERTIES OF MATTER

First Part. The excess pressure, $p = 2\gamma/r$, where γ is the surface tension and r is the radius of the bubble. This can be proved by considering the equilibrium of one-half of the bubble, when $\gamma \times 2\pi r = p \times \pi r^2$, leading to $p = 2\gamma/r$.

Second Part. If the atmospheric pressure is A and the pressure on the other side of the curved liquid surface is p_1, the excess pressure $= A - p_1$.

$$\therefore A - p_1 = \frac{2\gamma \cos 20°}{r} = \frac{2\gamma \cos 20°}{0.001} \quad \ldots \ldots \text{(i)}$$

since $r = 1$ mm $= 0.001$ m.

$$\text{Similarly, } A - p_2 = \frac{2\gamma \cos 20°}{0.0006} \quad \ldots \ldots \ldots \text{(ii)}$$

where p_2 is the pressure on the other side of the curved liquid surface. Subtracting (i) from (ii),

$$\therefore p_1 - p_2 = \frac{2\gamma \cos 20°}{0.0006} - \frac{2\gamma \cos 20°}{0.001}$$

$$= \frac{2 \times 0.05 \cos 20°}{0.0006} - \frac{2 \times 0.05 \cos 20°}{0.001} = 62.65$$

But $\quad p_1 - p_2 = h\rho g$,

where h is the difference in levels in m, ρ is the density in kg m^{-3} and g is 9.8 m s^{-2}.

$$\therefore h \times 800 \times 9.8 = 62.65$$

$$\therefore h = \frac{62.65}{800 \times 9.8} = 0.008 \text{ m}$$

3
Heat

11. HEAT CAPACITY. LATENT HEAT

(Where necessary, use specific heat capacity of water = 4200 J kg^{-1} K^{-1} and g = 10 N kg^{-1})

1 A heating coil in water has a p.d. of 40 V and carries a constant current of 2 A. Find (i) the heat supplied in 5 min, (ii) the temperature rise of the water in 5 min if its mass is 1 kg and the calorimeter has a mass of 0.2 kg and a specific heat capacity of 400 J kg^{-1} K^{-1}.

2 A lead bullet of mass 2 g, moving with a velocity of 20 m s^{-1}, is brought to rest at a target. Calculate the rise in temperature of the bullet, assuming all the heat produced is gained by it. (Specific heat capacity of lead = 130 J kg^{-1} K^{-1}.

3 The heights of two waterfalls are respectively 100 m and 50 m. Calculate the approximate difference in temperature between the top and bottom of each waterfall.

4 Calculate the cost of using 72 million joules of electrical energy if the cost is 3p per kilowatt-hour (kWh).

5 In the continuous-flow method, the p.d. across the heating coil is 40 V, the current in the coil is 2.5 A, the rate of flow is 3 g s^{-1}, and the steady temperature rise is 7°C. When the experiment is repeated with the rate of flow 1 g s^{-1}, the current is 1.4 A and the p.d. is 30 V when the steady rise in temperature is again 7°C. Calculate a value for the specific heat capacity of water from the results.

6 A car of 1000 kg is travelling at a speed of 72 km h^{-1}. The brakes have a mass of 20 kg. Assuming 20 per cent of the heat is developed at the brakes when they are used, calculate the rise in temperature of the brakes if the car is brought to rest. (Specific heat capacity of brake material = 500 J kg^{-1} K^{-1}.)

7 A skater moves round a circle of radius 10 m four times. If the frictional force at the ice is 100 N and each skate has a mass of 2 kg and a specific heat capacity of 500 J kg^{-1} K^{-1}, calculate the theoretical rise in temperature of each skate. Assume each skate is on the ground half the time.

8 A brass disc rotates steadily at 10 rev s^{-1} against a constant opposing couple due to friction of 20 N m. Calculate the work done after 10 s. If the disc has a mass of 2 kg and a specific heat capacity of 400 J kg^{-1} K^{-1}, and retains 80% of the mechanical work done as heat, calculate its temperature rise (*Note*. Work done = couple × angle in radians).

HEAT

9 An aluminium block, mass 0.5 kg is heated by a 40 W electrical heater embedded in it. In 1 minute, the temperature of the block rose from 15.0°C to 19.0°C. Calculate the specific heat capacity of aluminium, ignoring heat losses. If heat losses were *not* ignored, would the result be greater or less than your calculated value?

10 A metal block, mass 2 kg, is heated by a 60 W electrical heating coil until it reached a constant temperature. (i) What is the rate of loss of heat then lost to the surroundings? (ii) If the heating supply is cut off, and the metal is observed to cool initially at 4°C per minute, find the specific heat capacity of the metal.

11 State the meaning and units of *heat capacity* and *specific heat capacity* (c). 0.2 kg of iron at 100°C is dropped into 0.09 kg of water at 16°C inside a calorimeter of 0.15 kg and specific heat capacity 800 J kg^{-1} K^{-1}. Find the final temperature of the water (Sp.ht. capacity of iron = 400 J kg^{-1} K^{-1}.)

12 A copper calorimeter of mass 0.12 kg contains 0.1 kg of paraffin at 15°C. If 0.048 kg of aluminium at 100°C is transferred to the liquid, and the final temperature is 27°C, calculate the specific heat capacity of paraffin, neglecting heat losses. (Sp.ht. capacity of aluminium, copper = 1000, 400 J kg^{-1} K^{-1}.

13 In the continuous flow method, the p.d. across the heating coil is 1.4 V and the current is 2.6 A, and the mass of liquid flowing per second is 3.4 g; when conditions are steady the temperature rise is 2.0°C. When the rate of flow is decreased to 2.6 g per second and the p.d. and current are 0.9 V and 2.1 A respectively, the temperature rise is the same. Calculate the specific heat capacity of the liquid.

14 Describe the experiment to measure the specific heat capacity of a liquid by the *continuous flow* method. Draw a diagram of the apparatus, and give the theory of the method. What are the advantages of the method?

Latent Heat

15 What mass of iron at 17°C will cause 3.0 g of liquid oxygen at −183°C to evaporate if dropped into the liquid? (Specific latent heat of oxygen = 2.1 × 10^5 J kg^{-1}; sp.ht. capacity of iron = 400 J kg^{-1} K^{-1}; boiling point of liquid oxygen = −183°C.)

16 0.006 kg of steam at 100°C is passed into 0.08 kg of water at 14°C inside a copper calorimeter of mass 0.15 kg and specific heat capacity 400 J kg^{-1} K^{-1}. What is the final temperature? (Specific latent heat of steam = 2.3 × 10^6 J kg^{-1}.)

17 A heating coil immersed in water carries a current of 3.5 A and has a p.d. across it of 30 V. Calculate the mass of water evaporated in 10 min when boiling occurs, if the specific latent heat of steam is 2.25 × 10^6 J kg^{-1}. Describe an experiment, based on this method, of measuring the specific latent heat of steam.

18 0.02 kg of ice at $-15°C$ is placed inside 0.085 kg of water at $40°C$ in a 0.05 kg calorimeter of specific heat capacity 400 J $kg^{-1} K^{-1}$. Calculate the final temperature of the water. (l for ice = 3.4×10^5 J kg^{-1}; specific heat capacity of ice = 2000 J $kg^{-1} K^{-1}$.

19 A vacuum pump is connected to an insulated flask containing 50 g of water at $0°C$. As the water-vapour and air is pumped out, some of the water evaporates and the remaining water freezes at $0°C$. If the specific latent heat of evaporation of water at $0°C$ is 2.1×10^6 J kg^{-1}, and the specific latent heat of fusion of ice at $0°C = 3.4 \times 10^5$ J kg^{-1}, estimate the mass of water evaporated. (Assume the heat needed to evaporate the water is taken from the internal energy of the water.)

20 In measuring the specific latent heat of vaporization of ethyl alcohol by the electrical method, a mass of 4.2 g was collected in 5.0 min when the p.d. was 10.0 V and the current 1.5 A. When the p.d. was increased to 15.0 V and the current to 2.0 A, a mass of 9.3 g was collected in 5.0 min.

Calculate the specific latent heat of ethyl alcohol and the quantity of heat *per second* lost to the surroundings during the 5 min.

21 Draw a diagram of an apparatus for determining electrically the *specific latent heat of vaporization of steam*. List the errors in the experiment, and explain how you would minimize them.

22 Draw a diagram showing how the specific latent heat of ethyl alcohol may be measured by an electrical method, using a vapour jacket. Give the theory. List the advantages of the method.

12. EXPANSION OF GASES. IDEAL GAS EQUATION

1 A given mass of gas has a volume of 144 cm^3 at $15°C$. Calculate its volume at (i) $33°C$, (ii) $0°C$, (iii) $-67°C$, the pressure being constant.

2 The pressure of a given mass of gas at $27°C$ is 750.0 mmHg. If the volume remains constant, calculate the pressure at (i) $12°C$, (ii) $0°C$, (iii) $-50°C$.

3 A mass of gas has a volume of 22.0 litres at $19°C$ and pressure 1.1×10^5 N m^{-2}. What is the volume at $0°C$ and pressure 1.0×10^5 N m^{-2}?

4 A mass of gas has a volume of 30.0 litres at $17°C$ and 1.0×10^5 N m^{-2}. What is the pressure at $127°C$ when the volume is 25.0 litres?

5 Calculate the mass of air which has a volume of (i) 300 cm^3 at $15°C$ and 770 mmHg pressure, (ii) 1200 cm^3 at $100°C$ and 750 mmHg pressure (density of air = 1.29 kg m^{-3} at s.t.p.).

6 A barometer tube, 1 metre long above the outside mercury level, contains some air above the mercury inside it, and the height of the mercury inside stands 750 mm above the outside level of mercury. By how

HEAT

much is the tube depressed when the mercury inside is 745 mm above the outside level? (Assume the atmospheric pressure is 760 mmHg.)

7 A diving-bell of uniform cross-section is 3 m high, and contains air at 67°C when the atmospheric pressure is 770 mmHg pressure. The bell is then sunk into water at 17°C until the water rises 1 m inside it. Calculate the depth of the top of the bell below the water, assuming no air escapes.

8 A flask of 80 cm³ contains air at 17°C, and is stoppered with a capillary tube of bore 1 mm² cross-section containing a short mercury index. If the flask is warmed, calculate its new temperature when the pellet moves 120 mm along the capillary.

9 A gas at constant pressure has a volume of 500.0 cm³ at 20°C and a volume of 534.3 cm³ at 40°C. Calculate from these figures the absolute zero of the gas.

10 Calculate the mass of 225 cm³ of oxygen at 40°C and 770 mmHg pressure if the density of oxygen is 1.43 kg m⁻³ at s.t.p.

11 A piston pump of effective volume 150 cm³ is used to exhaust a vessel of volume (i) 1 litre, (ii) 2.5 litre. Calculate the number of complete strokes required to reduce the pressure to 0.01 atmosphere if the original pressure in the vessel is 1 atmosphere.

12 Calculate the molar gas constant if 1 mole of any gas occupies about 22.4 litres at 0°C and 760 mmHg. What mass of hydrogen has a gas constant of 100 J K⁻¹, and how many moles does it contain?

13 What is the *ideal gas equation*? Prove the relation between the pressure, volume and absolute (Kelvin) temperature of a *given mass* of gas, assuming the truth of Boyle's and Charles' laws.

14 Calculate the mass of argon in a vessel of volume 80 cm³ if the pressure is 750 mmHg and the temperature is 65°C. Density of argon = 1.56 kg m⁻³ at s.t.p.

15 Calculate the *gas constant for 1 kg* of (i) air, (ii) hydrogen. Density of air = 1.29 kg m³ at s.t.p.; density of hydrogen = 9 × 10⁻² kg m⁻³ at s.t.p. What is the gas constant for 5 kg of air?

16 For three different cases of an ideal gas, the equation of state may be expressed by (a) $pV = RT$, (b) $pV = nRT$, and (c) $pV = mRT/M$, where R is the molar gas constant and M is the mass of one mole.

(i) What is the number of moles used in (a), (b) and (c) respectively?
(ii) In (c), what does m represent?
(iii) Write down the equation of state for 1 kg of an ideal gas.
(iv) If N_A is the Avogadro constant, 6.0×10^{23} mol⁻¹, write down the equation of state for an ideal gas containing N molecules.

17 If the density of oxygen is 1.43 kg m⁻³ at 0°C and 1.013×10^5 N m⁻² pressure and the mass of one mole is 0.032 kg, calculate the molar gas constant.

18 In an experiment, 250 cm³ of helium gas is collected at 27°C and 750 mmHg pressure. Calculate the mass of this gas, given that the mass of one mole is 0.004 kg and the molar gas constant is 8.3 J mol^{-1} K^{-1}. (Assume 760 mmHg = 1.013 × 10^5 N m^{-2})

19 Calculate the number of moles in a gas which has a volume of 6 × 10^{-3} m^3 at a pressure of 10^5 N m^{-2} and temperature of 300 K. (Molar gas constant = 8.3 J mol^{-1} K^{-1}).

20 Two vessels at the same temperature are connected by a valve. Initially one vessel contains 3 litres of oxygen gas at 6.0 × 10^4 N m^{-2} pressure and the other contains 5 litres of nitrogen at 4.0 × 10^4 N m^{-2} pressure. Calculate the final equilibrium pressure.

21 Oxygen gas is contained in a vessel of volume 0.05 m^3 at a temperature of 17°C and pressure 2.0 × 10^5 N m^{-2}. When some of the gas is used, the pressure in the vessel falls to 0.8 × 10^5 N m^{-2} and the temperature remains 17°C. Calculate the mass of oxygen used. (Molar gas constant = 8.3 J mol^{-1} K^{-1}; mass of 1 mole oxygen = 0.032 kg.)

22 Write down a formula for the number of moles of a gas in terms of its pressure (p), volume (V), absolute temperature (T), and molar gas constant (R). A glass bulb of 100 cm^3 is connected to another bulb of 200 cm^3 by a narrow tube of negligible volume. The apparatus contains air at 17°C and 750 mm mercury pressure. The smaller bulb is then maintained at 57°C, the temperature of the other bulb remaining at 17°C. Calculate the new pressure in the apparatus.
(*Hint.* The total number of moles in the two bulbs remains constant.)

23 In Question 22, calculate the pressure when the smaller bulb is finally maintained at 100°C and the larger bulb at 27°C.

13. KINETIC THEORY OF GASES

1 List the fundamental assumptions of the kinetic theory of gases. A cube contains N molecules of a gas. On simple theory, how many molecules on the average may be assumed to move (i) along each of three perpendicular axes, (ii) in a direction towards, and perpendicular to, one particular face of the cube?

2 What is the difference between the *root-mean-square velocity* and the *mean velocity* of a group of gas molecules. The r.m.s. velocity is given by the expression $\sqrt{3p/\rho}$, where p is the gas pressure and ρ is the density. Show that this formula is dimensionally correct.

3 Calculate the r.m.s. velocity of hydrogen molecules at (i) 0°C, (ii) 27°C if the density of hydrogen is 0.09 kg m^{-3} at s.t.p.

4 Show how Boyle's law is explained by the kinetic theory of gases. State the assumptions made.

5 Derive an expression for the *root-mean-square velocity* of the molecules of a gas. Calculate the root-mean-square velocity of oxygen

HEAT

molecules at 0°C and at 100°C, if the density of oxygen at s.t.p. is 1.42 kg m^{-3}.

6 (i) The r.m.s. velocity of oxygen gas molecules is 500 m s^{-1} at a certain temperature and pressure. What would be the r.m.s. velocity of hydrogen gas molecules at the same temperature and pressure, if the molar mass of oxygen is 16 times that of hydrogen?

(ii) At what temperature will the r.m.s. velocity of hydrogen gas molecules be twice that at 300 K?

(iii) At what temperature will the r.m.s. velocity of nitrogen molecules equal that of oxygen molecules at 300 K, if the relative molecular masses of nitrogen and oxygen are 28 and 32 respectively?

7 Explain in terms of molecular theory:
(i) the *pressure* of a gas,
(ii) the increase in pressure when the volume of a gas is reduced at constant temperature,
(iii) the increase in pressure of a gas at constant volume when its temperature rises,
(iv) the rise in temperature when the air in the barrel of an air pump is compressed.

8 The mass of one mole of helium gas is 4 x 10^{-3} kg. Calculate the r.m.s. velocity of its molecules at s.t.p. if R = 8.3 J mol^{-1} K^{-1}.
What is the r.m.s. velocity of oxygen gas at s.t.p. if the mass of one mole of oxygen is 32 x 10^{-3} kg?

9 The pressure and heat capacity of a given mass of gas depends on the *mean square-speed* of all its molecules, whereas its rate of diffusion through a porous partition depends on the *mean speed* of all its molecules.
From the kinetic theory, account for this difference. Name another property of a gas which depends on its mean speed.

10 The number of molecules per unit volume in a given gas is 1.2 x 10^{24} m^{-3} at 27°C and 1.0 x 10^{5} N m^{-2} pressure. Calculate the number of molecules per unit volume of this gas at a temperature of −23°C and a pressure of 2.0 x 10^{3} N m^{-2}.

11 Derive the ideal gas equation $pV = RT$ from kinetic theory. State the assumptions made.

12 Hydrogen gas escapes much more quickly from a sealed container than chlorine gas due to leakage through a small hole. Explain why this occurs, given that the densities of hydrogen, air and chlorine are respectively 0.09, 1.2 and 3.2 kg m^{-3}.

13 The velocity of sound in air is given by $v = \sqrt{1.4p/\rho}$, where p is the pressure and ρ is the density of the air. Why is this relationship for v similar to that for the r.m.s. velocity of a gas, $v = \sqrt{3p/\rho}$?

14 (i) Why do the molecules in a gas have different speeds at a given instant?
(ii) Explain why one may assume, in a simple derivation of gas pressure,

that the average momentum change of molecules incident on a wall is constant in spite of collisions with other molecules.

15 State *Graham's law of diffusion of gases*. Deduce the law from the kinetic theory.

16 State *Avogadro's hypothesis*. How does the kinetic theory explain it?

14. HEAT CAPACITIES OF A GAS. ISOTHERMAL AND ADIABATIC CHANGES

(Assume molar gas constant $R = 8.3$ J mol^{-1} K^{-1})

Heat Capacities of Gas

1 Calculate the external work done by a gas if it expands from a volume of 0.02 m^3 to 0.08 m^3 under a constant pressure of 10^5 N m^{-2}.

2 (i) Define specific heat capacity of a gas at constant volume, c_V, and that at constant pressure, c_p. Which has the greater magnitude?

(ii) Define molar heat capacities C_V and C_p at constant volume and constant pressure respectively, and show that their difference is equal to R, the molar gas constant.

3 Assuming neon is an ideal monatomic gas of relative atomic mass 20, what are the values of (i) C_V, its molar heat capacity at constant volume, and (ii) c_V, its specific heat capacity at constant volume?

4 An ideal monatomic gas occupies a volume of 0.4 m^3 at a pressure of 10^6 N m^{-2} and temperature 300 K. Find (i) the number of moles of the gas, (ii) its internal energy.

5 The molar heat capacity of air at constant pressure is 29.1 J mol^{-1} K^{-1}. If 1 mole of air occupies 22.3×10^{-3} m^3 at $0°$C and 1.0×10^5 N m^{-2}, calculate a value for the molar heat capacity at constant volume.

6 For hydrogen, C_p and C_V are respectively 28.8 and 20.4 J mol^{-1} K^{-1}. Calculate a value for the molar gas constant and for the density of hydrogen at $0°$C and 1.013×10^5 N m^{-2} pressure if the mass of 1 mole is 2×10^{-3} kg.

7 Deduce a formula for the translational kinetic energy of 1 mole of an ideal monatomic gas. Hence deduce the molar heat capacities, C_V and C_p, of the gas and the ratio C_p/C_V.

Explain briefly why your expressions for C_V and C_p do not apply to an ideal *diatomic* gas.

8 (i) Assuming argon is an ideal monatomic gas of relative molecular mass 40, calculate c_V and c_p, its specific heat capacities at constant volume and constant pressure respectively.

(ii) 0.6 m^3 of argon has a density of 1.50 kg m^{-3} at a particular temperature and pressure. Calculate its heat capacity at constant volume.

HEAT

9 Helium is a monatomic gas of relative molecular mass 4 and has a density of 0.18 kg m^{-3} at s.t.p. ($0°C$ and 1.01325×10^5 N m^{-2} pressure). Find (i) the molar gas constant, (ii) the specific heat capacity at constant volume of helium, (iii) the internal energy of 1 kg of the gas at $27°C$.

10 Calculate the translational kinetic energy of the molecules of 4 kg of hydrogen at $27°C$ and 760 mmHg pressure. (Density of hydrogen at s.t.p. = 0.09 kg m^{-3}.)

11 Name three different forms of kinetic energy which a diatomic molecule may possess. Prove that the ratio of its molar heat capacities is 1.4 on the kinetic theory, stating any assumptions made.

12 1 g of water of volume 1 cm^3 forms 1671 cm^3 of steam on evaporation at $100°C$ and external pressure 1.013×10^5 N m^{-2}. If the specific latent heat of steam is 2.26×10^6 J kg^{-1}, calculate (i) the external work, (ii) the internal work done in this case.

13 0.01 kg of hydrogen is heated at constant volume from $15°C$ to $45°C$. What change is produced in the gas molecules? Calculate the increase in the internal energy of the gas if $c_V = 1.0 \times 10^4$ J kg^{-1} K^{-1}. (*Note.* c_V is the increase in internal energy per kg per K temperature rise.)

Isothermal and Adiabatic Changes

14 0.2 kg of a gas has a volume of 0.10 m^3 at $27°C$. It is heated at constant pressure of 1.5×10^5 N m^{-2} until its temperature rises to $87°C$. Calculate (i) the external work done, (ii) the heat supplied if $c_p = 1.0 \times 10^3$ J kg^{-1} K^{-1}, (iii) the increase in internal energy of the gas. (*Hint.* From $V \propto T$, calculate the new volume at $87°C$.)

15 Assuming helium is an ideal monatomic gas of relative molecular mass 4, estimate the internal energy of 1 kg of helium at 300 K. If the gas now doubles its volume by reversible adiabatic expansion, calculate the change in internal energy.

16 A gas expands *isothermally* and then *adiabatically*. (i) What does this statement mean? (ii) Draw a rough sketch of the pressure-volume variation of the two cases, and state the corresponding $p-V$ relation if the expansions are reversible.

17 A given mass of air contracts isothermally and reversibly from a volume of 100 cm^3 and pressure of 760 mmHg to a volume of 80 cm^3. Calculate the new pressure. What is the new pressure if the same change takes place adiabatically? ($\gamma = 1.4$ for air.)

18 A given mass of gas at $0°C$ has a pressure of 750 mmHg and a volume of 100 cm^3. Calculate (i) the new pressure, (ii) the new temperature, when the gas contracts adiabatically and reversibly to one half of its volume. ($\gamma = 1.4$).

19 Assuming the relation between the pressure and volume of a given gas under reversible adiabatic conditions, derive the relation between the temperature and the volume of the gas. The volume of a given mass of air is 600 cm³ when its temperature is 27°C. The gas undergoes an adiabatic expansion to 800 cm³. Calculate its new temperature. (γ = 1.4)

20 What information does γ provide about the molecules of a gas? Describe and explain one method of measuring γ.

21 The first law of thermodynamics may be expressed in the form $\delta Q = \delta U + \delta W$, where δQ is the heat supplied to a given system, δU is the internal energy change and δW is the external work done. State the respective changes of δQ, δU and δW in each of the following cases:-

(i) 1 mole of an ideal gas heated at constant volume through a temperature rise of 1 K,

(ii) 1 mole of an ideal gas heated at constant pressure through a temperature rise of 1 K,

(iii) 1 mole of an ideal gas expanding adiabatically and reversibly,

(iv) 1 mole of an ideal gas expanding isothermally and reversibly,

(v) m kg of water at 100°C changing to steam at 100°C,

(vi) m kg of ice at 0°C changing to water at 0°C.

15. VAPOURS. REAL GASES

Vapours

1 Define the terms *saturated vapour, unsaturated vapour*. Write down the laws which a saturated and an unsaturated vapour obey.

2 A closed vessel contains air and water at 7°C and the total pressure in the vessel is 780 mmHg. Calculate the new total pressure when the temperature rises to 27°C, water being still present in the vessel. (s.v.p. of water at 27°C, 7°C = 28, 12 mmHg pressure.)

3 Some air and water are present at the top of a column of mercury in a barometer. The total pressure is 80 mmHg and the temperature is 27°C. Calculate the s.v.p. of water at 60°C if the total pressure then becomes 210 mmHg, volume being constant. (s.v.p. of water at 27°C = 28 mmHg.)

4 Some air and saturated water-vapour are present in the space above the mercury inside a barometer tube. What happens to the total pressure when (i) the barometer tube is raised a little, (ii) the barometer is depressed a little, (iii) the temperature is raised, water remaining each time at the top of the tube? Give reasons.

5 Draw diagrams of the apparatus used to measure the variation of saturation vapour pressure of water in the respective ranges (i) 0°C–50°C, (ii) 50°C–100°C. Describe briefly the experiments.

6 A narrow tube of uniform bore closed at one end has some air trapped by a small pellet of water. The length of the column of air is 100 mm when the temperature is 14°C and the atmospheric pressure is 760 mm

HEAT 35

mercury. Calculate the length of the column when the temperature changes to 30°C. (s.v.p. of water at 14°C, 30°C = 12, 30 mmHg.)

7 Explain how the kinetic theory accounts for the phenomena of boiling, evaporation, and the variation of saturation vapour pressure with temperature.

8 A capillary tube of uniform bore, sealed at one end, contains some air trapped by a small pellet of water. At 18°C the length of the air column is 154 mm and at 52°C it is 226 mm long. If the atmospheric pressure is 760 mmHg, calculate the s.v.p. of water at 52°C. (s.v.p. at 18°C = 16 mmHg.)

9 Calculate the mass of 1 litre of air and water vapour at 30°C and 760 mmHg pressure if the water-vapour pressure is 11 mmHg. (The density of dry air is 1.18 kg m^{-3} at s.t.p., and the density of water-vapour is 5/8ths that of air at the same temperature and pressure.

10 A 1 litre flask contains dry air at 80°C and 750 mmHg pressure, the atmospheric pressure. The flask is opened in an inverted position under water at 15°C so that the water levels inside and out are the same. If the s.v.p. of water at 15°C is 16 mmHg pressure, calculate the volume of water present in the flask.

Real Gases

11 (i) Write down *two* assumptions made in the kinetic theory of gases which are not true for real gases. Are these assumptions more true for real gases at low or high pressures? Explain your answer.

(ii) Write down van der Waals' equation for a real gas. Why are the additional terms for p and V positive and negative respectively?

12 (i) What is *critical temperature*? Why is it important in liquefying gases?

(ii) Draw sketches showing the $p-V$ isothermals for a real gas above and below its critical temperature, and mark on the curves the liquid and gaseous states.

13 (i) Draw sketches of pV against p for (a) a real gas and (b) an ideal gas.

(ii) For a gas at constant temperature at lower pressures, $pV = A + Bp$ approximately, where A and B are constants. What is the significance of A?

14 Draw sketches showing the *critical isothermal* of a real gas for variations of p and V, an isothermal above the critical temperature, and an isothermal below the critical temperature. Explain the differences between the three curves.

At what part of the isothermals will the relation $pV = RT$ be roughly obeyed?

15 How do real gases differ from ideal gases? Write down *van der Waals'* equation, and explain how it was derived.

16. THERMAL CONDUCTION

1 The opposite faces of a solid metal cube are maintained at 100°C and 12°C respectively. The side of the cube is 0.1 m and its thermal conductivity is 250 W m^{-1} K^{-1}. Calculate (i) the temperature gradient in the cube, (ii) the quantity of heat flowing in 10 min through one face of the cube in the steady state.

2 Define *thermal conductivity*. What is a unit of thermal conductivity? Find the heat passing per hour through the brick walls of a room 4 m tall and whose floor is 5 m square, if the thickness of the bricks is 0.3 m the inside and outside surfaces are at 15°C and 5°C respectively. (Include the area of the door and windows in the area of the walls; $k = 0.4$ W m^{-1} K^{-1} for brick.)

3 Draw a labelled diagram of Searle's method of measuring the thermal conductivity of a good conductor. In the experiment, the temperatures at two points 0.1 m apart on a cylindrical metal bar are 72°C and 60°C respectively. If the diameter of the bar is 0.08 m, the mass of water collected in 30 s is 0.240 kg, and its temperature rise is 6.9°C, find the conductivity. ($c = 4200$ J kg^{-1} K^{-1} for water.)

4 A cylindrical bar of iron of length 10 cm is joined to a cylindrical bar of copper of length 15 cm and the same cross-sectional area, thus making a composite bar. If the ends of the iron and copper bars are maintained at 100°C and 10°C respectively, calculate the temperature of the junction of the bars. (Thermal conductivity of iron and copper = 50 and 380 W m^{-1} K^{-1}.)

5 The temperature of the water side of a metal boiler is 140°C and 3000 kg of water are evaporated per square metre of the boiler in 1 h. Calculate the temperature of the other side of the boiler if the thermal conductivity of the metal is 84 W m^{-1} K^{-1}, the boiler is 12 mm thick, and the specific latent heat of steam is 2.27×10^6 kg^{-1}.

6 What is meant by the statement that the thermal conductivity of brick is 48×10^{-3} W m^{-1} K^{-1}? A brick wall is 0.25 m thick and the temperature difference between the exposed surfaces is 30°C. Calculate the heat passing through it per square metre per hour. If the brick is covered on one side with plaster 5 cm thick, calculate the heat per square metre per hour now flowing, the temperature difference of the exposed surfaces being the same as before (thermal conductivity of plaster = 4×10^{-3} W m^{-1} K^{-1}).

7 One end of a metal bar is maintained at 100°C and the other end at 20°C when the bar is (i) lagged, (ii) unlagged. Draw sketches illustrating the temperature variation along the bar in each case, and explain the variation.

8 A composite bar is made of a bar of copper 10 cm long, a bar of iron 8 cm long, and a bar of aluminium 12 cm long, each having the same cross-sectional area. If the extreme ends of the bar are maintained at 100°C and 10°C respectively, find the temperatures at the two junctions. (Thermal conductivities of copper, iron, aluminium = 400, 40, 20 W m^{-1} K^{-1} respectively.)

9 A circular disc of glass, 3 mm thick and 110 mm diameter, is placed between two metal discs A, B in a 'Lees disc' experiment. The temperature of the lower disc, B, is constant at 93°C and the temperature of A at 96.5°C when steam is passed. When B is warmed above 93°C with the glass disc on it, and a cooling curve obtained, the rate of cooling at 93°C is found to be 0.042 K s^{-1}. Calculate the thermal conductivity of the glass if the mass of B is 0.94 kg and its specific heat capacity is 400 J kg^{-1} K^{-1}.

10 In experiments to measure the thermal conductivity of a bad solid conductor such as cardboard or ebonite, the substance is made thin and of fairly large surface area. Give reasons for choosing this shape. Describe an experiment to measure the thermal conductivity of cardboard, and give the theory of the method.

11 A glass tube of length 50 cm is totally immersed in a tank containing a large quantity of ice and water. Water at 45°C enters the tube and leaves at 15°C. Calculate the mass of ice melted per minute of the internal diameter of the tube is 10.2 mm, the external diameter is 10.6 mm, k for glass = 12.6 x 10^{-2} W m^{-1} K^{-1}, l of ice = 3.36 x 10^5 J kg^{-1}, c_w(water) = 4200 J kg^{-1} K^{-1}.

12 Describe an experiment to measure the thermal conductivity of a hollow glass tube. Give the theory of the method.

17. THERMAL RADIATION

(Where necessary, assume Stefan constant σ = 5.7 x 10^{-8} W m^{-2} K^{-4})

1 Draw a sketch of an instrument used to detect thermal radiation and explain its action. Describe an experiment which shows that a blackened surface is a good emitter of radiation and a polished silvered surface is a bad emitter.

2 What is *black-body radiation*? Draw rough sketches showing the variation with wavelength of the radiation from a black-body at temperatures of 1000 K and 3000 K respectively. Calculate the heat radiated per hour per metre^2 of a furnace at a temperature of 1727°C, assuming black-body radiation and that the temperature of the surroundings is negligibly small.

3 The sun's surface radiates energy at the rate of 6.2 x 10^7 W per metre^2 of its surface. Calculate the magnitude of the sun's temperature.

4 The intensity of the sun's radiation falling normally on the earth is about 1500 W per metre^2. If this radiation falls normally on a metal object of area 10 cm², mass 0.05 kg, and specific heat capacity 500 J kg⁻¹ K⁻¹, calculate the initial rate of temperature rise of the object? What happens to the temperature of the object as the sun's radiation continues to fall on it?

5 A 100 W lamp is connected to the mains at the specified voltage. When switched on, the filament soon reaches a constant absolute temperature T. (i) What is the rate of heat lost by the filament at this temperature? (ii) Assuming that the radiation from the filament is 0.8 times that from a black-body at the same temperature, estimate T using Stefan's law. (Surface area of filament = 5×10^{-5} m².)

6 A small hole in a large hollow sphere appears to be black when viewed in daylight, even though the interior may not be black. Explain this observation. Why may the hole be considered a 'black-body radiator'?

Draw a diagram of a laboratory apparatus you would use in a blackbody radiation experiment.

7 A solid metal sphere is placed in an enclosure of 27°C. When its temperature is 227°C it cools at the rate of 3.2 K per min. What is the rate of cooling when a solid sphere of the same metal but twice the radius, at 127°C, is placed in the same enclosure? Assume that Stefan's law applies.

8 Calculate the surface temperature of the sun from the following observations: Stefan constant = 5.7×10^{-8} W m⁻² K⁻⁴, solar constant (energy received normally per second from sun at earth's surface) = 1400 J m⁻²; sun's radius = 7×10^8 m; distance of sun from earth = 1.5×10^{11} m.

9 Draw a sketch showing how the energy E_λ in a narrow band of wavelengths of mean value λ, emitted by a black-body radiator at a constant temperature, varies with λ. In your diagram, show (i) the wavelength λ_m with maximum energy, (ii) that area of the graph which is a measure of the energy of a narrow band of wavelengths of mean value $\lambda_m/2$, (iii) that area which is a measure of the *total energy* emitted by the black-body.

10 Wien's law of radiation states that $\lambda_m \propto 1/T$. (i) Write down the meaning of the law, (ii) illustrate the law by drawing curves of E_λ v. λ at three different black body temperatures, (iii) show how the law is used to explain the colour changes in a hot metal object as its temperature rises.

11 Describe how the energy E_λ associated with a narrow band of different wave-lengths in the radiation from a hot object has been measured. What results were obtained? State *two* laws of radiation.

12 Draw a diagram of apparatus used to demonstrate that thermal radiation can be (*a*) reflected, (*b*) refracted. What evidence shows that the speed of light and of thermal radiation are the same?

HEAT

18. THERMAL EXPANSION OF SOLIDS AND LIQUIDS

Solids

1 The length of a steel rail at 10°C is 30.00 m. Find the new length in metres at (i) 100°C, (ii) 0°C. What is the *change* in length in millimetres when the temperature alters from 10°C to −10°C? (Mean linear expansivity of steel = $12 \times 10^{-6} K^{-1}$.)

2 The length of a simple pendulum of period 2.000 s at 8°C is 1 metre. Calculate the new period when the temperature changes to 18°C, assuming the suspending wire is iron of linear expansivity $12 \times 10^{-6} K^{-1}$.

3 A steel rod of length 10 m and diameter 20 mm at 10°C, is fixed to two rigid supports. Calculate the change in the tension in the rod when its temperature becomes (i) 110°C, (ii) 0°C. (Young's modulus for steel = $2 \times 10^{11} N m^{-2}$; linear expansivity of steel = $1.2 \times 10^{-5} K^{-1}$.)

4 The bob of a pendulum clock is suspended by a brass wire and keeps correct time, period 2 s, at 10°C. How much will the clock lose in a week if the temperature changes to 18°C? (Linear expansivity of brass = $1.9 \times 10^{-5} K^{-1}$.)

5 A steel girder of length 5 m and cross-sectional area 12 cm² is prevented from expanding when its temperature rises from 6°C to 13°C. Calculate the force exerted by the girder. (Young modulus for steel = $2 \times 10^{11} N m^{-2}$; linear expansivity of steel = $12 \times 10^{-6} K^{-1}$.)

6 The mean linear expansivity of brass is $18 \times 10^{-6} K^{-1}$. Write down the mean area (superficial) expansivity of brass, and the mean cubic expansivity. From first principles, prove the two formulae used.

7 An aluminium shed has an area of 250 m² at 10°C. Find the new area when the temperature changes to (i) −4°C, (ii) 30°C. (Mean linear expansivity of aluminium = $22 \times 10^{-6} K^{-1}$.)

Liquids

8 What is the relation between the densities of a liquid at 0°C and t °C in terms of its absolute cubic expansivity.

Mercury has a density of 13 590 kg m⁻³ at 0°C. Calculate the density at (i) 100°C, (ii) −10°C. (Absolute cubic expansivity of mercury = $18 \times 10^{-5} K^{-1}$.)

9 What is the advantage of Dulong and Petit's 'balancing column' method of measuring the absolute cubic expansivity of a liquid? If the height of the 'cold' column in an experiment is 678 mm at 15.75°C and the height of the 'hot' column is 709 mm at 100.1°C, calculate the absolute cubic expansivity of the liquid. Prove any formula you use.

10 The volume of mercury in a thermometer at 0°C is 0.62 cm³. Calculate the length of the capillary tube from 0°C to 100°C if its diameter

is 0.2 mm. (Mean apparent cubic expansivity of mercury = $18 \times 10^{-5} \mathrm{K}^{-1}$.)

11 Draw a diagram of a form of laboratory apparatus to measure the absolute cubic expansivity of mercury. Describe the experiment, and prove the formula for calculating the expansivity from the measurements taken.

12 A mercury thermometer has a glass tubing of internal bore 0.1 mm diameter, and the distance between the fixed points is 250 mm. Calculate the internal volume of the bulb and stem below the lower fixed point. (Linear expansivity of glass = $9 \times 10^{-6} \mathrm{K}^{-1}$; absolute cubic expansivity of mercury = $18 \times 10^{-5} \mathrm{K}^{-1}$.)

13 The height of a mercury barometer reads 758 mm on a steel scale at 16°C. What is the height 'corrected' to 0°C if the scale reads correctly at 0°C? (Linear expansivity of steel = $12 \times 10^{-6} \mathrm{K}^{-1}$; true cubic expansivity of mercury = $18 \times 10^{-5} \mathrm{K}^{-1}$.)

19. THERMOMETRY

(*Note.* (i) The thermodynamic (absolute) temperature is denoted by the symbol T and is measured in Kelvin, symbol K. (ii) The Celsius temperature is denoted by the symbol t and is measured in °C. (iii) $T = 273.15 + t = 273 + t$ (approx).)

1 On the thermodynamic temperature scale, the fixed point is the triple point of water (the temperature at which ice, water and water-vapour coexist in equilibrium) which is given the value 273.16 K. Using the relation $p_T/p_{tr} = T/273.16$ for measuring T with a constant volume gas thermometer, calculate the temperature in K when the pressure at an unknown temperature is (i) 800 mmHg, (ii) 560 mmHg, and the pressure at the triple point is 700 mmHg.

2 In a constant volume gas thermometer, the pressure at the ice point (0°C) is 600 mmHg and at an unknown temperature T it is 640 mmHg. Using $p_T/p_{ice} = T/273.15$, calculate T. Find the new temperature where the observed pressure is 520 mmHg.

3 The ice-point is 273.15 K on the thermodynamic scale. Calculate the thermodynamic temperature of a liquid if the resistance of a platinum resistance thermometer immersed in it is 3.800 Ω and at the ice point it is 3.500 Ω.

4 A mercury thermometer contains mercury with a volume of 0.500 cm³ at 0°C, and a volume of 0.520 cm³ at 100°C. What is the temperature on the Celsius scale when the volume of mercury is (i) 0.516, (ii) 0.524 cm³?

5 A platinum resistance thermometer has a resistance of 4.320 ohms at 0°C and a resistance of 4.860 ohms at 100°C. Calculate the temperature on the Celsius scale when the resistance is (i) 4.642, (ii) 4.210 ohms.

HEAT

6 The pressure of a gas in a constant volume gas thermometer is 680 mmHg at 0°C, and 929 mmHg at 100°C. Calculate the temperature in °C when the pressure is (i) 724 mmHg, (ii) 567 mmHg.

7 Draw a diagram of an accurate form of constant volume gas thermometer. Describe how you would measure the boiling point of salt solution with it.

8 Draw a diagram of a platinum resistance thermometer. Why are 'compensating leads' used? Show how the thermometer is connected to the special bridge circuit used. Describe how the temperature of a hot liquid is measured by this thermometer.

9 Name the range of temperature measured by (i) a mercury-in-glass thermometer, (ii) a platinum resistance thermometer, (iii) a constant-volume gas thermometer, (iv) a thermoelectric thermometer. Which of these four is (*a*) the most accurate thermometer, (*b*) the quickest to use, (*c*) the least accurate?

10 In a constant volume gas thermometer, the difference in levels of the mercury on either side is −100 mm at the temperature of melting ice, +140 mm at the temperature of steam at standard atmospheric pressure, and +30 mm at the temperature of a liquid. Calculate the liquid temperature in °C.

11 Describe the *thermoelectric thermometer*, and explain how it is used to measure the temperature of molten steel. What are the advantages and uses of this type of thermometer?

12 Draw a sketch of the principal features of an optical pyrometer *or* a radiation pyrometer. Explain the principles on which the pyrometer works. State one advantage and one disadvantage of the pyrometer you describe.

20. MULTIPLE CHOICE QUESTIONS – HEAT

For each of the questions 1–14, choose one statement from A to E which is the most appropriate.

1 In temperature measurement with various thermometers,
  A dummy leads are used with the platinum resistance thermometer to compensate for external electric fields,
  B the 'dead' space in the gas thermometer must be as small as possible,
  C the temperature on the optical pyrometer scale is read from a calibrated millivoltmeter,
  D the temperature must not be higher than 150°C if a mercury-in-glass thermometer is used,
  E air is prefered to nitrogen in an accurate form of gas thermometer.

2 In gases,
  A $C_p - C_V = R$ for ideal gases only,
  B $C_p - C_V = R$ for real gases,
  C the translational kinetic energy of a monatomic gas depends only on its volume,

D the internal energy of an ideal gas is independent of its temperature,
E the internal energy of an ideal gas is proportional to its temperature.

3 A slab of material X, 2 mm thick, and a slab of material Y, 20 mm thick and of the same area as X, are placed together in thermal contact. The outside face of X is maintained at 100°C, the outside face of Y at 0°C. If Y has 40 times the thermal conductivity of X, their common junction temperature is
  A 50°C
  B 40°C
  C 20°C
  D 10°C
  E 5°C

4 A black body radiator can be made by punching a small hole in a metal can having a lid. In this case
  A the inside of the can must be painted black,
  B the surface area round the hole is the actual black-body radiator,
  C the lid must be removed during a radiation experiment,
  D if the inside of the can is silvery, the can will act as a non-black-body radiator,
  E the inside of the can need not be painted black.

5 In the constant volume gas thermometer, if 'tr' represents the triple point of water, 't' the temperature in °C, and 'T' the temperature in K, then
  A $t = (p_t/p_o) \times 100$,
  B $T = (p_T/p_{tr}) \times 273.15$,
  C $t/100 = -(p_t - p_o)/p_o$ and $T/273.15 = p_{tr}/p_o$,
  D $T = (p_T/p_{tr}) \times 273.16$ and $t/100 = (p_t - p_o)/(p_{100} - p_o)$,
  E $T = p_T/373$ and $t = p_t/273$.

6 In the kinetic theory of gases,
  A the pressure is not affected by the attraction of the container walls on the molecules,
  B the mean pressure depends on the mean velocity of the molecules,
  C the pressure on a wall is independent of the time taken for molecules to travel to-and-fro across the container,
  D the force on the walls is proportional to the mean value of the momentum change,
  E the molecules all have the same velocity at a given instant.

7 The sun's surface radiates energy at the rate of about 6300 W cm^{-2}. If Stefan constant is 5.7×10^{-8} W m^{-2}K^{-4}, then if T is the absolute temperature of the sun's surface,
  A $5.7 \times 10^{-8} T^4 = 63 \times 10^6$,
  B $5.7 \times 10^{-8} T = 6300$,
  C $5.7 \times 10^{-8} T^4 = 6300$,
  D in 1 minute, the sun radiates $60 \times 5.7 \times 10^{-8} T^4$ joules,
  E the temperature in the heart of the sun is given by $5.7 \times 10^{-8} T = 63 \times 10^6$.

8 Solids such as glass are poor thermal conductors but metals are generally good conductors. This is because

HEAT

- A the molecules in glass do not move when heated,
- B the molecules in metals move more freely,
- C metals contain free electrons which transfer heat,
- D the molecules in glass vibrate through relatively small amplitudes when heated,
- E metals are chemically more active than insulators.

9 In reversible isothermal and adiabatic changes of an ideal gas,
- A there is a change in internal energy during isothermal expansion,
- B no work is done during adiabatic expansion,
- C no change in internal energy occurs during isothermal expansion,
- D the work done during adiabatic expansion is equal to the internal energy change,
- E isothermal curves are steeper than adiabatics at the points of intersection of the two sets of curves.

10 The total pressure of a mixture of air and saturated water-vapour at $17°C$ is 780 mmHg. The s.v.p. of water at $17°C$ is 8 mmHg. When the volume is kept constant and the temperature raised to $27°C$, the saturated vapour pressure is 12 mmHg. The total pressure p of the mixture is then calculated from
- A $p/300 = 780/290$,
- B $(p - 12)/300 = 772/290$,
- C $p = 780(1 + 10/273)$,
- D $p = (780 \times 27/17) + 12$,
- E $(p - 8) = 772 \times 27/17$.

11 In the Dulong and Petit method of measuring the absolute cubic expansivity of a liquid, one side of a U-vessel containing the liquid is heated to a high temperature. In this method
- A allowance must be made for the vessel expansion,
- B only the difference in levels of the liquid columns and their temperature difference are measured,
- C the density of the whole liquid remains unchanged,
- D the vessel diameter must be equal on both sides,
- E no allowance needs to be made for the expansion of the vessel.

12 Water evaporates into the vacuum at the top of a barometer tube and eventually becomes saturated. In this case,
- A the vapour now has minimum density,
- B the saturated vapour pressure depends on the volume at the top,
- C the vapour now has maximum density,
- D decreasing the volume at the top will increase the vapour pressure,
- E the vapour pressure depends on the atmospheric pressure,

13 Andrews' results for the isothermals of carbon dioxide show that
- A carbon dioxide can never be liquefied above a particular temperature,
- B carbon dioxide can always be liquefied above $31°C$,
- C the isothermals are S-shaped curves at all temperatures,
- D a liquid meniscus can be obtained above the critical temperature,
- E the pressure varies when the substance changes from the liquid to the gaseous state.

14 In the flow tube method of measuring the specific heat capacity, c

of a liquid,
A a fixed volume of liquid is always used,
B electrical heating produces a large temperature difference between the inflow and outflow of the liquid,
C the heat lost per second can never be eliminated,
D c is most accurately found by increasing the current and speed of flow to produce the same temperatures as before,
E the experiment can not be used to find the variation of c with temperature change.

15 Parts (i) to (vii) below are statements about gases. The relationship of these statements to the simple kinetic theory of gases is required. Answer:
A if the statement is false and does not follow from kinetic theory
B if the statement is true but has nothing to do with the kinetic theory of gases
C if the statement is true and lends support to the kinetic theory of gases
D if the statement is true but contradicts the simple kinetic theory of gases
E if the statement is false but is a prediction of the simple kinetic theory of gases
Statements —
(i) The pressure of a fixed mass of an ideal gas is inversely proportional to the volume if the temperature is constant.
(ii) Gases can only be liquified when below their critical temperature
(iii) All ideal gases at the same temperature and pressure contain the same number of molecules
(iv) The molar heat capacity of an ideal monatomic gas at constant volume is $5R/2$
(v) Gases are normally colourless
(vi) The ratio γ, of the molar heat capacities of a diatomic gas is 1.4
(vii) The force of attraction between molecules is important at high pressures.

16 Give the number of degrees of freedom in the following cases (a)–(e). Choose your answers from
A 0, B 2, C 3, D 5, E 6.
(a) Rotational degrees of a diatomic gas at room temperature
(b) Translational degrees of a monatomic gas
(c) Translational degrees of a diatomic gas
(d) Total degrees of a polyatomic gas at room temperature
(e) Rotational degrees of a monatomic gas at room temperature

17 One mole of an ideal monatomic gas is at 300 K and 10^5 N m^{-2} pressure. It is compressed to half its volume at constant pressure. Take R as 8.3 J mol^{-1} K^{-1} so that the volume of the gas is 25×10^{-3} m^3 initially. Calculate
(a) The work done on the gas:
A −1250 J, B 1250 J, C 625 J, D −625 J, E −3112.5 J
(b) The heat supplied to the gas:

A 4560 J, B −4560 J, C 3537.5 J, D −3537.5 J, E None of these.
(c) The change in internal energy:
A −1862.5 J, B 1862.5 J, C 3410 J, D −3410 J, E None of these.
18 The graph shows the cooling curve for a material (Fig. 20A).

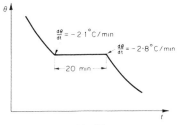

Fig. 20A

The specific heat capacity of the solid material is 1.2 kJ kg^{-1} K^{-1}.
(a) The specific heat capacity of the liquid material in kJ kg^{-1} K^{-1} is
A 1.9, B 0.9, C 1.2, D 1.6, E None of these.
(b) The specific latent heat of the substance in kJ kg^{-1} is
A 89.6, B 67.2, C 37.8, D 50.4, E None of these.

19 The following types of thermometer are to be used for various purposes:
A Mercury-in-glass, B Platinum thermometer, C Thermocouple,
D Hydrogen gas thermometer, E Pyrometer.
Which are most suitable for the following purposes:-
(a) Measuring the temperatures across a small crystal to determine its thermal conductivity
(b) Plotting a cooling curve for sulphur liquid
(c) Measuring the temperature of a furnace
(d) Calibrating another thermometer
(e) In a constant flow calorimeter
(f) Measuring the temperature distribution along a copper bar
(g) The temperature of a chemical reaction taking place in a sealed container.

20 The graph shows the temperature distribution along a bar of linear expansivity 10^{-5} K^{-1}. (Fig. 20B).

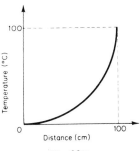

Fig. 20B

The total expansion is

A 5×10^{-2} cm, B 10^{-1} cm, C 2×10^{-2} cm, D 2.5×10^{-2} cm
E 3.5×10^{-2} cm.

21 The $p - V$ graphs (Fig. 20C) show two isotherms for fluid(s) contained in a cylinder. According to conditions the cylinder may contain

A a gas,
B a gas and a vapour,
C a gas, liquid and a vapour,
D a liquid,
E a gas and liquid.

Which gives the contents of the cylinder at (*a*) position (i), (*b*) position (ii), (*c*) position (iii), (*d*) position (iv)?

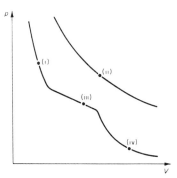

Fig. 20C.

WORKED EXAMPLES ON HEAT

1 A lead bullet of 0.025 kg enters a target with a velocity of 150 m s^{-1} and is brought to rest in the target. Assuming three-quarters of the heat produced is absorbed by the bullet, calculate the temperature rise of the bullet. (Specific heat capacity of lead = 130 J kg^{-1} K^{-1}.)

$$\text{Energy of bullet} = \tfrac{1}{2} mv^2 = \tfrac{1}{2} \times 0.025 \times 150^2 \text{ J}$$

$$= 281.25 \text{ J} = \text{heat produced}$$

$$\therefore \quad 0.025 \times 130 \times \theta = \tfrac{3}{4} \times 281.25,$$

where θ is the temperature rise of the bullet.

$$\therefore \quad \theta = \frac{3 \times 281.25}{4 \times 0.025 \times 130}$$

$$= 65 \text{ K}$$

2 The density of air is 1.29 kg m^{-3} at s.t.p. Calculate (i) the molar gas constant assuming the mass of 1 mole of air is 28.8×10^{-3} kg and (ii) the root-mean-square velocity of air molecules at s.t.p.

(i) We have $p = h\rho g = 0.76 \times 13\,600 \times 9.8$ N m^{-2}, since $p = 760$ mmHg, and volume of 28.8×10^{-3} kg = $28.8 \times 10^{-3}/1.29$ m^3.

HEAT

$$\therefore \quad R = \frac{pV}{T} = \frac{0.76 \times 13\,600 \times 9.8 \times 28.8 \times 10^{-3}}{273 \times 1.29}$$

$$= 8.3 \text{ J mol}^{-1} \text{K}^{-1}$$

(ii) r.m.s. velocity $\sqrt{\overline{c^2}} = \sqrt{\frac{3p}{\rho}}$

$$= \sqrt{\frac{3 \times 0.76 \times 13\,600 \times 9.8}{1.29}}$$

$$= 485 \text{ m s}^{-1}$$

3 A flask of 200 cm³ is joined to a flask of 100 cm³ by a narrow tube, and each flask is at 17°C. The air in the flask has a pressure of 750 mmHg. What is the new pressure if the temperature of the larger flask is altered to 77°C, the temperature of the smaller flask being unaltered?

The total mass of air in the flasks is constant. Now $pV = mrT$, where m is the mass of a gas and r is the gas constant per unit mass for air.

$$\therefore \quad \text{mass, } m, = \frac{pV}{rT}$$

$$\therefore \quad \text{total mass originally} = \frac{750 \times 200}{r \times 290} + \frac{750 \times 100}{r \times 290}.$$

$$\text{Also, total mass finally} = \frac{p \times 200}{r \times 350} + \frac{p \times 100}{r \times 290},$$

where p is the new pressure, since the absolute temperature of the 200 cm³ flask is now (273 + 77) or 350 K.

$$\therefore \quad \frac{p \times 200}{r \times 350} + \frac{p \times 100}{r \times 290} = \frac{750 \times 200}{r \times 290} + \frac{750 \times 100}{r \times 290}.$$

Cancelling r throughout, and then simplifying, we find

$$p = 850 \text{ mmHg}$$

(*Note.* In a calculation of the actual mass of gas, we would need to change 750 mmHg pressure to N m⁻², and 200 and 100 cm³ to m³. In the above problem, however, the additional factors needed would cancel on both sides of the equation and may therefore be omitted in this particular case.)

4 Calculate the heat conducted per metre² per min through (i) a brick wall 10 cm thick, (ii) through the same wall lined on one side with cork 2 cm thick, the temperature difference between the exposed surfaces being 20°C in both cases. (Thermal conductivity of brick and cork = 0.48 and 0.04 W m⁻¹ K⁻¹ respectively.)

$$Q = kAt \times \text{temperature gradient,}$$

where Q is the quantity of heat in the steady state passing through a material of thermal conductivity k and cross-sectional area A in a time t.

(i) For the brick wall alone, $A = 1$ m^2, $t = 60$ s, temperature gradient $= 20/0.1 = 200$ K m^{-1}, and $k = 0.48$ W m^{-1} K^{-1}.

$$\therefore \quad Q = 0.48 \times 1 \times 60 \times 200 = 5760 \text{ J}.$$

(ii) With the cork on one side, we must first find the temperature excess of the *junction* of the brick and cork over the outside temperatures. Now in the steady state, the quantity of heat per second per metre^2 passing through the cork and brick is the same. Since the temperature gradient for the cork $= (20 - \theta)/0.02$, and that for the brick $= \theta/0.1$, where θ is the junction temperature.

$$\therefore \quad 0.04 \times \frac{20 - \theta}{0.02} = 0.48 \times \frac{\theta}{0.1}$$

Solving
$$\therefore \quad \theta = 5.9°\text{C}.$$

The quantity of heat, Q, passing per square metre through the brick in 1 min is thus given by

$$Q = kAt \times \text{temperature gradient}$$

$$= 0.48 \times 1^2 \times 60 \times \frac{5.9}{0.1}$$

$$= 1700 \text{ J}.$$

5 The density of steam is 0.6 kg m^{-3}. Calculate the heat required to do (i) work against the atmospheric pressure (external work), (ii) internal work, when 1 kg of water at 100°C is changed to 1 kg of steam at 100°C. (Assume atmospheric pressure $= 1.013 \times 10^5$ N m^{-2}, $l = 2250$ kJ kg^{-1}.)

The volume of 1 kg of water $= 10^{-3}$ m^3.

The volume of 1 kg of steam $= \frac{1}{0.6}$ m$^3 = 1666.7 \times 10^{-3}$ m^3. Thus the change in volume from water to steam $= (1666.7 - 1) \times 10^{-3} = 1665.7 \times 10^{-3}$ m^3.

(i) Now external work in joule (J)

$$= \text{pressure in N m}^{-2} \times \text{volume change in m}^3$$

$$= 1.013 \times 10^5 \times 1665.7 \times 10^{-3} \text{ J}$$

$$\therefore \quad \text{heat} = 1.013 \times 10^5 \times 1665.7 \times 10^{-3} \text{ J} = 168 \text{ kJ}.$$

(ii) Total heat required $= 2250$ kJ

$$\therefore \quad \text{heat used for internal work} = 2250 - 168 = 2082 \text{ kJ}$$

6 What is the difference between an *isothermal* and an *adiabatic* change? An ideal gas at 750 mmHg pressure is compressed isothermally until its volume is reduced to three-quarters of its original value. It is then allowed to expand adiabatically to a volume 20 per cent greater than its original volume. If the initial temperature of the gas is 17°C, calculate its final pressure and temperature. ($c_p = 3600$, $c_V = 2400$ J kg^{-1} K^{-1} for the gas.)

HEAT 49

First Part. An isothermal change is one which takes place at constant temperature. An adiabatic change is one which takes place at constant heat, i.e., no heat enters or leaves the system.

Second Part. The ratio of the specific heats, $\gamma = \dfrac{c_p}{c_V} = \dfrac{3600}{2400} = 1.5$.

Under isothermal conditions, Boyle's law is obeyed, i.e., pV = constant,

$\therefore \quad p \times \tfrac{3}{4} V = 750 \times V$, where p is the new pressure.

$\therefore \quad p = 1000$ mmHg.

Under adiabatic conditions, pV^γ = constant. Hence the new pressure p is given by

$$p \times (\tfrac{120}{100} V)^\gamma = 1000 \times (\tfrac{3}{4} V)^\gamma$$

$$\therefore \quad p = 1000 \times \left(\dfrac{\tfrac{3}{4}V}{\tfrac{6}{5}V}\right)^{1.5}$$

$$= 1000 \times (\tfrac{5}{8})^{1.5} = 494 \text{ mmHg}.$$

7 A piece of metal of mass 0.5 g at $18°C$ is dropped into liquid oxygen at its boiling-point $-183°C$. Some of the liquid vaporizes, producing 130 cm^3 of oxygen at $18°C$ and 75 cm mercury pressure. Calculate the mean specific heat capacity of the metal if the specific latent heat of vaporization of liquid oxygen is 2.14×10^5 J kg^{-1}, and the density of oxygen gas is 1.43 kg m^{-3} at s.t.p.

The volume of oxygen at s.t.p. $= 130 \times \dfrac{273}{291} \times \dfrac{75}{76} = 120.3$ cm^3

$\therefore \quad$ mass of oxygen $= 1.43 \times \dfrac{120.3}{10^6} = 1.7 \times 10^{-4}$ kg

$\therefore \quad$ heat produced by copper $= ml = 1.7 \times 10^{-4} \times 2.14 \times 10^5$ J

$\therefore \quad 0.5 \times 10^{-3} \times c \times [18 - (-183)] = 1.7 \times 10^{-4} \times 2.14 \times 10^5$

$\therefore \quad c = 360$ J kg^{-1} K^{-1}

8 Distinguish between a *saturated* and an *unsaturated* vapour. A uniform capillary tube is sealed at one end, and contains some air enclosed by a small water pellet. At $18°C$ the length of the air column is 146 mm. At $60°C$ the length of the air column is 228 mm. If the saturation vapour-pressure of water at $18°C$ is 15 mmHg, calculate its saturation vapour-pressure at $60°C$. (Barometric height = 755 mmHg.)

First Part. A saturated vapour is one in equilibrium with its liquid. An unsaturated vapour is one not in equilibrium with its liquid.

Second Part. The pressure, p_1, of air in the capillary tube at $18°C =$ $755 - 15 = 740$ mmHg, from Dalton's law of partial pressure, since the total pressure of air and water-vapour = atmospheric pressure.

The pressure, p_2, of the air in the tube at $60°C = (755 - p)$ mmHg, where p is the saturation vapour pressure of water vapour at $60°C$. Now $pV/T =$ constant, for the given mass of air.

$$\therefore \frac{(755-p) \times 228}{273+60} = \frac{740 \times 146}{273+18}$$

$$\therefore 755 - p = \frac{740 \times 146 \times 333}{228 \times 291}$$

$$\therefore p = 213 \text{ mmHg}.$$

9 The height of a mercury barometer is read at $18°C$ as 756.0 mm on a brass scale correct at $0°C$. Calculate the height of mercury at $0°C$ exerting the same pressure. (Cubic expansivity of mercury = $182 \times 10^{-6} \text{ K}^{-1}$; linear expansivity of brass = $18 \times 10^{-6} \text{ K}^{-1}$.)

Since the brass expands from $0°C$ to $18°C$, the true height of the column at $°C$, h_t, is given by

$$h_t = 756(1+\alpha t) = 756(1 + 0.000018 \times 18).$$

Suppose h_0 is the height at $0°C$, and ρ_0, ρ_t are the densities of mercury at $0°C$, $t°C$.

Then, since pressure = $h\rho g$, $h_0\rho_0 g = h_t\rho_t g$

$$\therefore h_0 = h_t \frac{\rho_t}{\rho_0} = h_t \times \frac{1}{1+0.000182 \times 18}$$

$$= \frac{756(1+0.000018 \times 18)}{1+0.000182 \times 18} = 756[1 + 18(0000018 - 0.000182)]$$

approx,

$= 753.8$ mm.

4
Geometrical Optics

21. REFLECTION AT MIRRORS

Plane Mirrors

1 Two mirrors A, B are inclined to each other at an angle of 60°. Draw all the images obtained when a point object is placed between the mirrors so that the angles made by the mirrors and the line joining the object to their line of intersection are 20°, 40° respectively. Show also the rays by which an observer sees the third image when looking into mirror A.

2 A man 2.0 m tall stands in front of a plane mirror so that he can *just* see the top of his head and his feet. Calculate the length of the mirror.

3 Prove that, when the direction of the incident ray is constant, the angle of rotation of a mirror is half the deviation of the reflected ray.

4 Two plane mirrors are inclined at 30° to each other. Show (i) by drawing, (ii) by calculation that the angle of deviation produced by successive reflection at both mirrors is 60° when the ray is incident at 50° on one mirror. Repeat for an angle of incidence of 40°.

5 A point object is placed between two parallel plane mirrors 6 cm apart. After how many successive reflections is the distance of the image from the object 36 cm?

6 Two plane mirrors are inclined to each other at a fixed angle θ. Show that the angle of deviation of a ray which undergoes two successive reflections at the mirrors is always 2θ, no matter what angle of incidence the ray makes with the first mirror.

7 Describe (i) a periscope, (ii) a sextant. Explain fully the optical principles of each instrument.

Spherical Mirrors

8 An object is placed (i) 6 cm, (ii) 3 cm from a concave mirror of focal length 5 cm. Find the image distance from the mirror and the magnification.

9 The image in a concave mirror is real, and twice as long as the object. If the radius of curvature of the mirror is 20 cm, calculate the object distance from the mirror. Draw a ray diagram showing how the image is formed.

10 A convex mirror has a focal length of 18 cm. Draw roughly the image of an object (i) 22.5 cm, (ii) 9 cm from the mirror. Find by calculation the image distance from the mirror and the magnification in each case.

11 An image in a convex mirror of radius 16 cm is one-third as long as the object. Calculate the object distance from the mirror.

12 The image in a concave mirror is upright and five times as tall as the object. If the focal length of the mirror is 25 cm, calculate the image distance from the mirror. Draw also a sketch.

13 An object is placed on the axis of a concave mirror so that first a virtual, then a real image, each three times as long as the object, is formed. Find the distance between the two images if the radius of curvature of the mirror is b.

14 A pole 6 m long is laid along the axis of a convex mirror of radius of curvature 3 m, the end of the pole nearer the mirror being 2 m from the mirror. Calculate the length of the pole seen in the mirror.

15 An object is magnified three times by a concave mirror, and the image is received on a screen. The object and the screen are now moved until the image is four times the size of the object. If the screen has then been moved 30 cm, calculate the focal length of the mirror and the shift of the object.

16 Describe and explain an accurate method of measuring the focal length of (i) a concave mirror, (ii) a convex mirror.

22. REFRACTION AT PLANE SURFACES

1 Define the *refractive index* from (i) air to glass, (ii) glass to water, (iii) water to air in terms of angles, drawing a sketch in each case. What are these three refractive index values in terms of the velocity of light?

2 A ray of light is incident at $35°$ (i) from air to glass, (ii) from glass to air, (iii) from water to glass. Find the angle of refraction in each case. ($n_g = 1.5, n_w = 1.33$.)

3 A travelling microscope reads 6.540, 8.380, 11.425 cm when focused respectively on an object in air, when a rectangular glass slab is placed on the object and the microscope re-focused, and when the microscope is focused on top of the glass. Calculate the refractive index of the glass. Prove the formula you use.

4 A vessel contains a slab of glass 10 cm thick. Calculate the apparent displacement of an object at the bottom of the slab when viewed directly above the glass. What is the displacement of the object if a layer of water 6 cm thick is placed on the glass and the object is viewed through both substances? ($n_g = 1\frac{1}{2}, n_w = 1\frac{1}{3}$.)

5 An object is placed below a parallel-sided block of glass and viewed directly above the glass. Prove that the apparent displacement of the object is independent of its position below the glass.

6 Describe how the refractive index of a small quantity of liquid can be measured by means of a concave mirror. Give the theory of the method, stating clearly the assumptions made.

GEOMETRICAL OPTICS 53

7 Calculate the critical angle between the following media: (i) air-glass, (ii) air-water, (iii) glass-water. Draw sketches showing the critical angle in each case. ($n_g = 1.51, n_w = 1.33$.)

8 Describe fully how the refractive index of a liquid can be measured by the *air-cell method*. Give the theory of the method. Why is white light not used for accuracy?

9 State the conditions under which total internal reflection occurs. Using diagrams, explain (i) why an isosceles right-angled prism acts as a good reflector of light, (ii) how a mirage occurs.

10 A rectangular glass slab 9 cm thick has a layer of water 6 cm thick on it, and a layer of oil 4 cm thick is on the water. If the refractive indices of glass, water and oil are 1.5, 1.33, 1.1 respectively, calculate the apparent displacement of an object beneath the glass slab when viewed from above.

11 A point object is placed 12 cm from a concave mirror of radius of curvature 8 cm. Calculate the displacement of the image position when a small rectangular piece of glass, 3 cm thick, is placed perpendicular to the axis between the object and the mirror. ($n_g = 1.5$.)

23. REFRACTION THROUGH PRISMS

1 The angle of a glass prism is 60°. Calculate the angle of emergence into the air of a ray incident at 45° on one face of the prism. ($n_g = 1.5$.)

2 The angle of minimum deviation produced by a prism of 60° is 42°30′. Calculate the refractive index of the glass.

3 How are the telescope and collimator of the spectrometer adjusted? The refractive index of a 60° glass prism is 1.52. Calculate the minimum deviation produced by the prism.

4 Derive the formula for the refractive index of a prism in terms of its angle A, and the angle of minimum deviation, D.

5 Calculate the least angle of incidence on a 60° glass prism of refractive index 1.5 when all the light at the second face is totally reflected.

6 Draw diagrams of (i) a spectrometer, (ii) the spectrometer arrangement when the angle of the prism, and the minimum deviation, are respectively measured. Describe how the angle of the prism and the minimum deviation are both measured.

7 (i) Draw a graph showing how the deviation of a ray by a prism varies with the angle of incidence on one face; on the graph mark two angles of incidence which give maximum deviation. (ii) Describe, with the aid of diagrams, how the table of the spectrometer is levelled.

8 A prism has angles of 90°, 45°, 45°. A ray parallel to the largest (hypotenuse) face enters the prism, and is refracted towards this face. Prove that total reflection takes place at the latter face. (Refractive index of glass = 1.5.)

9 A ray is incident on a prism of angle 70° so that it just grazes the surface. Calculate the angle of emergence of the ray from the prism if its refractive index is 1.64.

10 A prism of 60° has a refractive index of 1.52. Calculate the smallest angle of incidence on the prism for the ray just to emerge from the opposite face. What is the deviation of the ray?

11 Calculate the largest angle of a glass prism for a ray to emerge from the prism after refraction through the opposite face, if the refractive index is 1.52.

12 A ray of light incident on a 70° glass prism emerges in air after refraction at both sides, making an angle of 60° with the normal. Calculate the angle of incidence at the first face of the prism. ($n_g = 1.6$)

24. REFRACTION BY LENSES

1 An object is placed (i) 15 cm (ii) 6 cm from a converging lens of focal length 12 cm. Calculate the image distance from the lens in each case. Draw ray diagrams to show how the images are formed.

2 The image in a converging lens of focal length 20 cm is three times as long as the object. Calculate the two possible object distances from the lens, and draw sketches to illustrate.

3 Find the distance of the image from a diverging lens of focal length 10 cm if the object is (i) 20 cm, (ii) 5 cm from the lens. Draw ray sketches to illustrate.

4 The image in a diverging lens of focal length 30 cm is one-quarter of the object length. Calculate the object distance from the lens.

5 When a converging lens is used as a magnifying glass the object is magnified eight times. If the distance between the image and object is 14 cm, find the focal length of the lens.

6 Calculate the focal length of (i) a plano-concave glass lens of radius 10 cm and refractive index 1.5, (ii) a bi-convex lens of radii 15 cm and 30 cm respectively and refractive index 1.6.

7 A beam of light converges to a point 18 cm behind (i) a converging lens of focal length 12 cm, (ii) a diverging lens of focal length 36 cm. Calculate in each case the image distance from the lens, and draw sketches to illustrate.

8 Describe a method of finding the focal length of a converging lens mounted in a tube so that no measurements can be made to its surfaces. Prove the formula used.

9 A diverging lens is placed between an illuminated object and a concave mirror so that an image is obtained beside the object. The distance from the object to the lens is 20 cm and the distance from the lens to the mirror is 12 cm. Calculate the focal length of the lens if the radius of curvature of the mirror is 15 cm.

GEOMETRICAL OPTICS

10 A converging lens is moved between a slit and a screen so that two clear images are obtained on the screen. In one case the width of the slit is 6.4 mm and in the other case it is 2.5 mm. Calculate the actual width of the slit, and explain your calculations.

11 Prove that an image cannot be received on a screen with a converging lens unless the distance between the object and image is equal to, or greater than, four times the focal length.

12 Describe two different methods of measuring the focal length of (i) a converging lens, (ii) a diverging lens. Give the theory in each case.

13 An object is placed 15 cm from a converging lens of focal length 10 cm, and a diverging lens of focal length 8 cm is placed 24 cm from the converging lens on the other side. Find the distance of the final image from the diverging lens.

14 An illuminated object is placed 60 cm from a concave mirror of radius 55 cm. A diverging lens of focal length 10 cm is placed between them so that an image is obtained beside the object. Calculate the object distance from the lens.

15 Prove that the minimum distance between an object and a real image for a converging lens is $4f$, where f is the focal length. Describe an experiment to measure the focal length of a converging lens based on this result.

16 Describe, giving the necessary theory, how you would find the refractive index of a small quantity of liquid with the aid of a converging lens and plane mirror.

17 Describe how you would measure the radii of curvature of the surfaces of (i) a diverging, (ii) a converging lens, by an optical method.

An illuminated object is placed 8 cm in front of a converging lens of focal length 12 cm. An image is then obtained beside the object by reflection from the *back* surface of the lens. Calculate the radius of curvature of the back surface.

18 A thin bi-convex lens has a refractive index 1.5 and surfaces of radii of curvature 12 cm and 8 cm respectively; a thin bi-concave lens has a refractive index 1.5 and surfaces of radii 10 cm and 15 cm respectively. Calculate the combined focal length when the two lenses are placed in contact.

19 Show that, in general, a converging lens produces two images of an object on a screen when the object and screen are fixed. The ratio of the lengths of the two images is 6.25, and the distance between the object and screen is 100 cm. Calculate the focal length of the lens and the distance between the two positions of the lens.

20 A thin liquid lens is formed between a bi-convex lens of focal length 10 cm and a plane mirror. The focal length of the combination is found to be 16.0 cm; when the lens is turned over the focal length of the combination is 16.5 cm. Calculate the refractive index of the liquid if the refractive index of the glass is 1.5.

21 A vessel contains water to a depth of 350 mm. A converging lens of focal length 210 mm is immersed just below the water surface. Calculate the distance from the water-top of the image of an object at the bottom of the water, when viewed from above, ($n_g = 3/2, n_w = 4/3$.)

22 (i) Define the *dioptre,* symbol D. A lens has a power of +2D. What do you know about the lens? (ii) The power of a lens is +5D. Calculate the focal length of a lens which, when combined with the first lens, produces a lens of +3D.

Defects of Vision

23 An elderly person cannot see clearly objects nearer than 250 cm. What spectacles will he need to reduce the distance to 25 cm? Find the range of distinct vision if he can focus rays converging to points not less than 100 cm behind his eyes.

24 A person can see objects clearly at distances from 20 cm to 150 cm from his eyes. What spectacles are required to enable him to see distant objects clearly? What is his least distance of distinct vision when using the spectacles?

25 A short-sighted man has a range of vision from 16 cm to 24 cm from the eye. What lens should be used in order to enable him to see distant objects clearly, and what is the range of accommodation when using the lens?

26 A man uses reading spectacles consisting of a converging lens of 40 cm focal length for his left eye, and a diverging lens of 30 cm for his right eye. If the normal reading distance is 25 cm, what deductions do you make about the man's sight? Calculate his near point for each eye.

27 A man observes objects clearly when they are 15 cm to 300 cm from the eye. What lens enables him to see distant objects clearly? Calculate his least distance of distinct vision when using the lens.

25. DISPERSION BY PRISMS AND LENSES

In the questions, assume that $n_b = 1.54, n_r = 1.52, n$ (mean) $= 1.53$ for CROWN glass, and $n_b = 1.66, n_r = 1.62, n$ (mean) $= 1.64$ for FLINT glass, unless otherwise is specified.

Prisms

1 State the relation for the deviation of a ray of light by a small-angle prism when the angle of incidence is small.

A ray of white light is incident on a crown glass 6° prism at a small angle of incidence. Calculate the mean (yellow light) deviation and the dispersion between the blue and red.

GEOMETRICAL OPTICS

2 Repeat Question 1 if the prism is a flint glass 4° prism.

3 Calculate the dispersive power of crown glass and of flint glass using their given values of refractive index.

4 Calculate the angle of a flint glass prism which makes an achromatic combination with a crown glass prism of 8°. What is the mean deviation of the light by the combination? Draw a sketch showing the emergent red and blue rays.

5 A direct-vision spectroscope is made with a crown glass and a flint glass prism. Find the angle of the crown glass prism if the flint glass prism is 5°. What is the dispersion between the red and blue rays produced by the combination? Draw a sketch showing the emergent mean (yellow) light.

6 Derive the relation between the angles of a flint glass and crown glass small-angle prisms when they produce (i) no dispersion between the red and blue light, (ii) no deviation of the mean light.

Lenses

7 What is *chromatic aberration*? Show with the aid of a diagram the chromatic aberration produced by a converging lens, with a parallel beam of white light incident on it.

8 The radii of curvature of a crown glass converging lens are 20.0 cm and 30.0 cm respectively. Calculate the focal length of the lens for red and for blue light. If a parallel beam of white light is incident on the lens, describe the appearance of the colours on a screen as it is moved through the focus of the lens.

9 A flint glass lens has a mean focal length of 50.0 cm (i.e., for yellow light). Calculate the focal length for blue and for red light.

10 Draw a diagram of an *achromatic telescope objective,* and explain briefly why it is achromatic. Why cannot a converging and diverging lens of the same material be joined together to form an achromatic combination?

11 Write down the condition for two lenses to form an achromatic doublet. A crown glass converging lens of focal length 80 cm is combined with a flint glass lens to form an achromatic doublet. Calculate the focal length of the flint lens and of the combination.

12 Explain why an object appears uncoloured if it is viewed through a converging lens acting as a magnifying glass.

13 A white disc of diameter 1.00 cm is placed 40 cm in front of a thin converging crown glass lens of mean focal length 20 cm. Calculate the diameters of the red and blue images of the disc, and their distance apart.

14 An achromatic converging lens of 200 cm focal length is made by joining together a flint and crown glass lens. Calculate the focal lengths of the two lenses, and state which is diverging.

26. OPTICAL INSTRUMENTS

Telescopes

1 What is meant by the term *visual angle*? Find the ratio of the visual angles when an object is placed 30 cm and then 90 cm from the eye.

2 Define the *angular magnification* (*magnifying power*) of a telescope in normal use. An astronomical telescope consists of converging lenses of 30 cm and 5 cm focal length respectively. Calculate the angular magnification from first principles when the final image is formed (i) at infinity, (ii) 25 cm from the eye placed close to the eyepiece. Draw diagrams of the ray paths in each case.

3 An astronomical telescope in normal use has an angular magnification of 40. If the objective focal length is 100 cm, calculate the eyepiece focal length and the distance apart of the lenses. Find the angular magnification when the eyepiece is moved so that the final image is formed 25 cm from the eye.

4 An astronomical telescope objective has a focal length of 1 metre and diameter 80 mm. The angular magnification required is x 20. Calculate (i) the focal length of the eyepiece, (ii) its least diameter if all the light transmitted through the objective from a distant point on the axis is required to fall on the eyepiece, and the telescope is in normal adjustment.

5 The largest telescope in the world is a *reflecting telescope*. What are *two* advantages of this type of telescope? Explain with ray diagrams how the image is viewed with the aid of a small convex mirror.

6 Where is the *eye-ring* in the case of an astronomical telescope? Prove that the magnifying power of an astronomical telescope in normal adjustment is equal to the ratio of the objective and eye-ring diameters.

7 A distant luminous object subtends an angle of $0.5°$ at the centre of the objective of an astronomical telescope in normal adjustment. The objective of the telescope has a focal length of 80 cm, the eyepiece has a focal length of 14 cm. The light emerging from the eyepiece is focused by a camera lens of $f = 20$ cm. Calculate the length of the image on the photographic plate.

Microscopes

8 A converging lens of 4 cm focal length is placed close to the eye. Calculate the magnifying power if the magnified erect image is formed (i) at infinity, (ii) 25 cm from the eye.

9 Define the angular magnification of a *microscope* in normal use. The objective and eyepiece of a microscope are converging lenses of 2 cm and 4 cm focal length respectively, and an object is placed 3 cm in front of the

objective. If the final image is formed 25 cm from the eye, calculate the distance between the lenses and the angular magnification of the instrument. Draw a diagram of the ray paths.

10 Repeat Question 9 if the converging lenses are 2.5 cm and 5 cm focal length respectively.

11 A microscope has an objective of 4 cm focal length and an eyepiece of 5 cm focal length. The distance apart of the lenses is 30 cm, and the final image is formed 25 cm from the eye. Calculate the distance of the object from the objective lens and the magnifying power of the microscope. Draw a ray diagram showing the formation of the image.

12 A microscope has an objective of 5 cm focal length and an eyepiece of 7.5 cm focal length. An object is placed 5.5 cm from the objective, and the final image is formed 25 cm from the eye close to the eyepiece. Find the distance between the lenses and the magnifying power of the microscope.

Other Instruments

13 An opera glass has a converging lens and diverging lens of 40 cm and 5 cm respectively. Find the magnifying power when it is focused on a distant object and the final image is 25 cm from the eye. Calculate the eye-ring distance from the eyepiece and show the eye-ring in a diagram.

14 Draw a diagram of the objective, eyepiece and the two prisms in one half of a *prism binoculars*. What are the purposes of the two prisms? What are the advantages over opera glasses? If the focal lengths of the objective and eyepiece are 32 and 4 cm respectively, what is the approximate distance between the objective and eyepiece?

15 Draw a diagram of the essential components of a *slide projector*, and draw a ray diagram showing the formation of the image on the screen. The distance from the lens to the screen is 7.5 m and the screen is 2.4 m square. If the slides used are 10 cm square, calculate the focal length of the most suitable projection lens.

16 A slide projector has a lens of focal length 30 cm. What range of movement must be provided for focusing the lens if the distance from projector to screen is required to vary from 18 m to 12 m? What is the ratio of the magnification produced at the distances 18 m and 12 m?

17 What is meant by the *f*-numbers '*f*-4' and '*f*-8' for a camera lens? If the exposure time for a *f*-8 setting is 1/64 s, what is the exposure time for a *f*-4 setting?

18 Define *relative aperture* of a lens. If d is the diameter of the lens aperture of a camera lens and f is the focal length, show that the brightness of the image formed is proportional to d^2/f^2. Hence explain why the sequences of *f*-numbers is *f*-2.8, 3.5, 5.6.

5
Physical Optics

27. WAVE THEORY OF LIGHT. VELOCITY OF LIGHT

(Assume the velocity of light *in vacuo* (or air), $c = 3.0 \times 10^8$ m s^{-1})

Wave Theory

1 Define *refractive index* in terms of the velocity of light. Calculate the wavelength in glass, refractive index 1.5, of (i) violet light of wavelength 4.5×10^{-7} m in air, (ii) red light of wavelength 7.0×10^{-7} m in air. What is the physical nature of the vibrations of light?

2 State *Huygens' Principle*. Using the principle, and showing many of the wavelets, draw sketches of wavefronts illustrating how a parallel beam of light is reflected from a plane mirror. Prove the law of reflection.

3 Plane wavefronts travelling in the direction of the principal axis are incident on (i) a concave spherical mirror and (ii) a convex spherical mirror each of radius r.
Draw sketches showing the incident and reflected wavefronts in each case. Mark on your sketches the positions of the centre of curvature and the principal focus, and explain briefly why $f = r/2$ from consideration of the wavefronts.

4 The refractive index of a sample of glass is 1.643 for blue light, and 1.618 for red light. Calculate the difference in the velocity of light in the glass for the two colours.

5 A particular type of glass has refractive indices for blue and red light which are $n_b = 15.2$ and $n_r = 1.50$ respectively. Calculate the speeds of blue and red light in the glass. From your results, which colour is deviated more when a parallel beam of white light is refracted into the glass?

6 A parallel beam of white light is incident on an air-glass surface. Draw sketches of the refracted wavefronts for blue and red light respectively to illustrate how dispersion occurs.

7 Explain Snell's law of refraction on the wave theory. What experimental evidence led to the rejection of Newton's corpuscular theory?

8 A point object is in front of a plane mirror. Show, by the wave theory, that its image is the same distance behind the mirror.

9 The wavelength in air of yellow light is 6.0×10^{-7} m. Calculate its wavelength in water of refractive index 4/3.

10 Calculate the thickness of a block of glass, refractive index 3/2, which has the same number of wavelengths as 18 cm of water, refractive index 4/3, when each is traversed by the same monochromatic light.

PHYSICAL OPTICS

11 By means of the wave theory, explain the phenomenon of total internal reflection. Derive a formula for the critical angle in terms of the speed of light.

12 A plane wavefront is incident at minimum deviation on a 60° glass prism. (i) Draw a sketch showing how the wavefront is refracted symmetrically through the prism and into the air. (ii) By considering the equal times taken for the light to travel through the glass base of the prism and through the air at the top, derive the formula $n = \sin(A + D)/2 \div \sin A/2$ for the refractive index of the glass, where A is the angle of the prism and D is the minimum deviation.

Velocity of Light

13 In Fizeau's method for determining the speed of light, the rotating wheel had 720 teeth, and the distance between the wheel and reflector was 8633 m. If the number of revolutions per second when extinction of light first occurred was 12.6, calculate (i) the time taken for the light to travel from the wheel to the reflector and back, (ii) the distance moved by the wheel in this time, (iii) the speed of light. What is seen as the rate of revolution of the wheel is increased to twice this speed and to three times the speed?

14 In Foucault's rotating mirror method for determining the speed of light, a plane mirror A is rotated at 250 rev s^{-1}. A beam of light is reflected from A towards a fixed mirror 10 km away, and on arriving back at A, the beam is reflected by A so that it makes an angle of 12° with its original direction. Calculate the speed of light from these measurements.

15 Draw a diagram of the rotating wheel (Fizeau) method of measuring the speed of light. Give the theory of the method. What is a disadvantage of the method?

16 In a Michelson method of measuring the speed of light c, an octagonal prism is rotated with increasing speed until the point source is first seen again. If this occurs at a rate of revolution of 940 rev s^{-1}, and the light travels a total distance of 40 km before returning to the prism, estimate c.

17 In a form of laboratory apparatus, the speed of light is measured by means of a beam of light reflected from a plane mirror M towards a spherical concave mirror 2 metres away. It then travels back to M, where it is reflected to a graduated scale S 1.6 m from M. When M is rotated at 400 rev s^{-1}, the observed deflection of the beam along the scale is 0.11 mm. Estimate the speed of light from the measurements.

18 Describe fully Michelson's method of measuring the velocity of light. List the advantages over the rotating mirror method.

28. INTERFERENCE

1 In a Young's experiment, the separation of the bright bands with monochromatic light is 0.54 mm, the separation of the slits is 1.0 mm, and the distance of the slits from the focal plane of the eyepiece observing the band is 1.00 m. Calculate the wavelength of the light.

2 Using the data in Question 1, what is the distance apart of the bright bands for violet light of wavelength 4.4×10^{-7} m and for red light of wavelength 7.5×10^{-7} m in Young's experiment. Describe the appearance of the bands when the slits are illuminated by a source of white light.

3 Draw a diagram of the experiment to measure wavelength by Young's interference experiment. Show the region in which interference of the light beams occur, and state the formula for the measurement of wavelength.

4 In Young's experiment, describe and explain the effect on the appearance of the bands of (i) moving the source slit nearer to the two slits, (ii) moving the two slits together, (iii) moving the screen further away, (iv) replacing a monochromatic source of red light by one of blue light.

5 Describe and explain how interference bands are produced in a narrow air-wedge film, mentioning (i) the coherent sources, (ii) where the bands are located, (iii) the explanation of the dark band where the plates touch.

6 Two flat microscope slides are in contact along one edge, and the opposite edges are separated by a thin piece of foil. When the air-wedge formed is illuminated normally by a sodium flame and the reflected light is viewed, 64 bright interference bands are counted from one end of the wedge to the other. Calculate the thickness of the foil, and explain how the bands are formed. (Wavelength of sodium light = 5.89×10^{-7} m.)

7 In question **6**, the spacing of the bright bands is 0.20 mm. Find (i) the angle in radians of the air wedge, (ii) the new spacing when light of wavelength 4.73×10^{-7} m is used.

8 Describe and explain the formation of *Newton's rings*. In Newton's rings experiment, the diameter of the 20th bright ring 67.8 mm and the radius of the lower surface of the lens is 1.00 m. Calculate a value for the wavelength of the light used.

9 In a Newton's rings experiment, a phase change occurs in the incident waves.
(i) Where does the phase change occur?
(ii) What effect does this produce on the optical path?
(iii) Give a reason for this phase change and describe briefly how you would demonstrate that your reason was correct.

10 Draw a diagram of *Lloyd's mirror* experiment for interference bands. In the diagram show (i) the coherent sources, (ii) the beams which overlap and produce interference bands. State the formula for wavelength in the experiment.

PHYSICAL OPTICS 63

11 A lens produces a series of Newton's rings. If the diameter of the 15th bright ring is 9.6 mm and the wavelength of the light used is 6.0×10^{-7} m, calculate the radius of the lower lens surface.

12 Interference bands are formed 0.30 mm apart in a small air-wedge film when monochromatic light of wavelength 6.0×10^{-7} m is used. What is the effect on the bands of (i) raising one plate parallel to itself through a very small distance of 6×10^{-3} mm (ii) replacing the air by a liquid of refractive index 4/3, (iii) using monochromatic light of wavelength 5.0×10^{-7} m?

13 Draw a diagram of Newton's rings experiment. Show in the diagram (i) the coherent sources which produce the rings, (ii) where the rings are situated, (iii) the appearance of the central spot. How does the appearance of the rings change when they are viewed by transmitted light?

14 In a Newton's rings experiment, the diameter of the nth ring is 5.60 mm and the diameter of the $(n + 10)$th ring is 8.00 mm. Calculate the wavelength of the light used if the lower lens surface has a radius of curvature of 1.50 m.

29. DIFFRACTION

1 Monochromatic light of wavelength 5.0×10^{-7} m falls normally on a diffraction grating having 600 lines mm^{-1}. Calculate the angular position of the first order spectrum on one side of the normal. What is the highest order of spectrum obtained on this side of the normal?

2 A diffraction grating has 800 lines mm^{-1} and is illuminated normally by parallel monochromatic light of wavelengths 5.6×10^{-7} and 5.9×10^{-7} m respectively. Calculate the difference in angular positions of the first order spectra on one side of the normal. What is the distance between the two first order spectra on a photographic plate if a camera lens of focal length 0.50 m is pointed directly at them?

3 A plane transmission diffraction grating has 600 lines mm^{-1} and a second order image is obtained at an angle of $43°20'$ when the grating is illuminated normally by monochromatic light. Calculate the wavelength of the light. What is the angular position of the first order image?

4 Two plane transmission gratings have 200 and 3000 lines mm^{-1} respectively. Describe qualitatively the diffraction images obtained when they are illuminated normally by (i) monochromatic light, (ii) white light.

5 Describe how a spectrometer and plane diffraction grating are used to measure the wavelength of sodium light. State particularly the necessary adjustments of the spectrometer table for the diffraction grating and give the reasons for the adjustments.

6 A diffraction grating has 400 line mm^{-1} and is illuminated normally by monochromatic light of wavelengths 4.5×10^{-7} and 6.0×10^{-7} m. In what respective order of spectrum does each colour overlap, and what is the corresponding angle of diffraction?

7 Describe the appearance on a screen if a very narrow slit is illuminated normally by a parallel beam of monochromatic light.

8 The *resolving power* of a telescope is found from the relation $\theta = 1.22\lambda/D$.
(i) Explain the meaning of this relation.
(ii) What is the reason for the factor '1.22'?
(iii) Does this relation apply to a reflector *or* refractor *or* radio telescope *or* to all three types of telescopes?
(iv) Is the resolving power of a telescope affected by the focal length or the diameter of the eyepiece used?

9 Calculate the resolving power (limit of resolution) for (i) the Yerkes Observatory telescope, which has an objective lens of 1.00 m, (ii) the 5.00 m Mount Palomar parabolic reflector telescope. Assume a mean wavelength of 6.0×10^{-7} m for white light.

10 (i) State two advantages of a large objective for telescopes, (ii) calculate the resolving power of the Jodrell Bank 75 m radio-telescope 'bowl' for radio waves of wavelength 200 mm.

11 Two telescopes X and Y in normal adjustment have the following properties:

	X	Y
objective diameter	10 cm	40 cm
objective f	0.8 m	1.6 m
eyepiece f	0.05 m	0.1 m

For X and Y, what is the ratio of (i) their angular magnifications (magnifying powers), (ii) their angular resolution, (iii) the brightness of the images formed their objectives?

12 A compound microscope has a resolving power given by '$\lambda/2n \sin \alpha$'.
(i) Explain the meaning of λ, n and α, drawing a diagram, to illustrate.
(ii) Name two ways of increasing the resolving power of a microscope,
(iii) What type of microscope produces the highest resolving power, and what is the reason for such high resolving power?

30. POLARIZATION

1 What is *plane-polarized light*? How does it differ from ordinary (unpolarized) light? Draw a sketch to illustrate your answer. What type of wave is a light wave?

2 Light reflected from the surface of a desk is observed through a Polaroid, which is rotated. Describe and explain what is seen.

PHYSICAL OPTICS

3 What is *Brewster's law*? Calculate the polarizing angle for glass of refractive index 1.52.
Draw a sketch showing the incident, reflected and transmitted rays when light is incident at the polarizing angle on the glass.

4 What is meant by *double refraction*? Describe, and explain, what is observed when a block of calcite is placed on an ink spot on a piece of paper and then rotated.

5 Describe the Nicol prism, and explain why it produces plane-polarized light. How would you use the prism to test whether a beam of light was plane-polarized?

6 What is a *longitudinal wave* and a *transverse wave*? Describe and explain experiments which show that light is a transverse wave.

7 Explain why the image of a lamp, seen by reflection in glass at an angle of about 56°, fluctuates in intensity when observed through a rotating sheet of Polaroid.
Explain the use of polarized light in *saccharimetry* or in *stress analysis*.

31. MULTIPLE CHOICE QUESTIONS – OPTICS

For each of the questions **1–10**, *choose one statement from* **A** *to* **E** *which is the most appropriate.*

1 In the case of a diverging lens,
   A images are always real,
   B images of real objects may sometimes be magnified,
   C the sun's rays can be brought to a focus,
   D a virtual object may produce a real image,
   E both faces of the lens must always be concave.

2 The angle of a glass prism is 60° and the minimum deviation is 40°.
   A The refractive index of the glass is the ratio 60° : 40°,
   B the maximum deviation is 100°,
   C the angle of incidence at minimum deviation is 50°,
   D there are two values of angle of incidence at minimum deviation,
   E the angle of incidence at minimum deviation is 90°.

3 A biconvex lens has equal radii of curvature of 20 cm and a refractive index of 1.5. The focal length is
   A 20 cm, **B** 10 cm, **C** 40 cm, **D** 30 cm, **E** 21.5 cm.

4 In a compound microscope with an objective O and eyepiece E,
   A the object is placed 25 cm from the eye,
   B the angular magnification is the ratio of the visual angles when image and object are both 25 cm from the eye,
   C the angular magnification is f_O/f_E,
   D O is a converging lens and E is a diverging lens,
   E the object is placed nearer O than its focal length.

5 In Young's interference experiment with monochromatic light,
   A the bright and dark bands are due to dispersion,
   B the energy lost in the dark bands is converted to heat,

66 GRADED EXERCISES AND WORKED EXAMPLES IN PHYSICS

    C the path difference for successive bright bands is $m\lambda/2$ where m is an integer,
    D only a point source can be used,
    E the energy lost in the dark bands is present in the bright bands.

6 Two cameras have lenses X, Y respectively with apertures of equal diameters and focal lengths of 5 cm and 8 cm respectively,
    A the camera with lens X will require the longer exposure time,
    B the illumination per unit area of the image produced by X and Y are in the ratio 5 : 8 respectively,
    C the camera with lens Y will require the longer exposure time,
    D the illumination per unit area of the image produced by X and Y are in the ratio 8 : 5 respectively,
    E the exposure times needed are the same in both cases.

7 The two images of an object viewed through calcite spar are due to
    A light waves polarized in perpendicular directions,
    B total internal reflection,
    C light waves polarized in parallel directions,
    D partial absorption of some of the incident light,
    E light waves polarized by reflection.

8 A condenser in a projection lantern is used
    A to produce enlarged images,
    B to provide high illumination of the object,
    C to project the image forward,
    D to reduce the illumination of the image,
    E to reduce colour effect.

9 Images are formed by a diffraction grating when monochromatic light is incident normally on it. If the grating is now reversed so that the light is incident normally on the opposite face
    A no images are obtained,
    B the images are displaced further on one side,
    C the images are produced in the same angular positions,
    D the angular positions of the images are reduced,
    E only light incident normally on the grating is diffracted.

10 In the case of an astronomical telescope in normal adjustment,
    A the angular magnification is the ratio of the diameters of the objective and eyepiece,
    B the final image is at the least distance of distinct vision,
    C an eyepiece of long focal length increases the angular magnification,
    D a real image is produced beyond the eyepiece when this lens is pulled out,
    E the final image is real.

11 (Fig. 31A).
    When the eye is moved to the observer's right the object appears

Fig. 31A.

PHYSICAL OPTICS

to the left of the image, when the eye is moved to the left the object appears to the right of the image. No parallax may be obtained by
A moving the mirror away from the lens,
B moving the mirror towards the lens,
C moving the pin away from the lens,
D moving the pin towards the lens,
E moving the pin away from the lens and the mirror towards the lens.

12 The following graphs A to E refer to the variance of certain quantities in geometrical optics, (the graphs cover real and virtual object distances, unless otherwise stated).

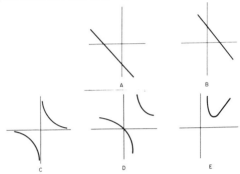

Fig. 31B.

Which graph will result from the following plots? (a) $1/u$ against $1/v$ for a diverging lens, (b) $u + v$ against u for real images in a converging lens, (c) $1/u$ against $1/v$ for a converging lens, (d) u against v for a converging lens, (e) $1/u$ against $1/v$ for a convex mirror, (f) $(u - f)$ against $(v - f)$ for a converging lens.

13 An object in a rectangular vat of clear liquid appears 2 m down in the liquid when viewed obliquely from above. The real depth of the tank is 3 m. Which of the following statements is true?
A the refractive index of the liquid is 1.5,
B the refractive index of the liquid is less than 1.5,
C the object appears at the correct distance from the side,
D the object appears to be at a greater distance from the side than it really is,
E none of these.

14 (Fig. 31C).

Fig. 31C.

The diagram shows the usual arrangement for showing Young's fringes. The following changes to the fringe pattern may be observed:

A The fringes get further apart,
B the fringes get closer together,
C the fringes get fainter and further apart,
D the fringes get brighter and closer together,
E the fringes get brighter but they become blurred.

Which of these describes best what happens in the following cases?
(a) The screen T is moved to the left,
(b) the screen T is moved to the right,
(c) the distance between the slits L and M is increased slightly,
(d) the distance between the slits L and M is decreased slightly,
(e) the slit S is widened,
(f) the screen containing the slits L and M is moved to the right,
(g) the screen containing the slits L and M is moved to the left while the screen T is moved to the left by the same amount.

Instructions

In the following *Assertion–Reason* questions, answer
A if the assertion, the reason and the explanations are all true,
B if the assertion and reason are both true but the explanation is false in the sense that it does NOT fit the assertion,
C if the assertion is true and the reason is false,
D if the assertion is false and the reason is true,
E if the assertion and reason are both false.

Assertion	*Reason*
15 Colours are seen in soap bubbles	Light is diffracted through soap solution
16 The bands in an air wedge film are closer together where the gap is most narrow	The optical path difference changes uniformly with distance along the wedge
17 Newton's rings have a dark spot in the centre	Phase change of π at a denser medium
18 Red light is deviated more than blue light by a glass prism	Red light travels slower than blue light in glass
19 When white light is incident normally on a diffraction grating, the first order spectrum is coloured red nearer the incident direction and blue on the other side	The first order spectrum is given by $d \sin \theta = \lambda$
20 A telescope with a large objective resolves distant objects well	The angular resolving power is given by $\theta = 1.22 \lambda/d$
21 Prism binoculars have a wide field of view	The prisms turn the image round and make it erect
22 The galilean telescope produces an erect image	This is due to the converging lens used as the objective

PHYSICAL OPTICS

	Assertion	Reason
23	The first order image in a coarse diffraction grating is bright	The dispersion of the grating is usually low
24	Light waves are longitudinal waves	Light waves are electromagnetic waves
25	When a grating of 500 lines per millimetre is illuminated normally by monochromatic light of wavelength 6×10^{-7} m, only one diffraction image is obtained.	$\sin i / \sin r$ = constant
26	Light reflected from a glass surface of refractive index $n = 1.5$ has maximum polarization when the angle of incidence is $57°$	$\tan i = n$
27	X-rays cannot be transmitted through crystals	X-rays are diffracted by crystals
28	Radio waves cannot be polarized	Radio waves are longitudinal waves
29	Sound waves cannot be diffracted	Sound waves are longitudinal waves
30	Illumination by ultraviolet increases the resolving power of a microscope	Resolving power is proportional to the wavelength
31	Two images are seen through calcite spar	Double reflection occurs
32	Colours are seen in oil films	White light consists of numerous colours
33	POLAROID filters absorb all the unpolarized light incident on them	Unpolarized light cannot pass through crystalline media
34	X-rays cannot be polarized	X-rays are electromagnetic waves

WORKED EXAMPLES ON OPTICS

1 In a Young's slits experiment, the distance from the slits to the screen is 1.00 m, the wavelength of the light used is 6×10^{-7} m, and the distance from the centre of the fringe pattern to the 10th bright band on one side is 30 mm. Calculate the separation of the slits.

A film of transparent material is placed over one of the slits and the displacement of the centre of the fringe pattern was observed to be 30 mm. Calculate the refractive index of the material if its thickness is 2×10^{-2} mm.

If a is the separation of the slits, D is the distance from the slits to the screen and y is the separation between successive bright bands, then, since

$$\frac{\lambda}{a} = \frac{y}{D},$$

$$\therefore a = \frac{\lambda D}{y} = \frac{6 \times 10^{-7} \times 1}{3 \times 10^{-3}} = 2 \times 10^{-4} \text{ m} = 0.2 \text{ mm}.$$

The displacement of the fringe pattern = 30 mm. This corresponds to a path difference of 10λ, from the first part of the question. If t is the film thickness and n its refractive index, the extra path difference = $nt - t = (n-1)t$.

$$\therefore (n-1) \times 2 \times 10^{-5} = 10 \times 6 \times 10^{-7}$$

$$\therefore n = 1 + \frac{6 \times 10^{-6}}{2 \times 10^{-5}} = 1.3$$

2 Define the terms: *angular magnification of a telescope, near point, far point.*

An astronomical telescope has converging lenses of focal lengths 100 cm, 5 cm respectively, and the image of a distant object is formed 25 cm from the eyepiece. (i) Calculate the distance between the lenses and the angular magnification if the eye is close to the eyepiece. (ii) What is the best position of the eye?

First Part. The 'angular magnification', M, = α'/α, where α', α are the angles subtended at the eye (visual angles) by the final image and object respectively. The 'near point' is the nearest point to the eye at which an object can be focused clearly; the 'far-point' is the further point from the eye at which an object can be focused clearly ('infinity' for normal vision).

Second Part. (i) The image I_1 in the objective lens O is formed 100 cm from O, since rays from the distant object arrive parallel at the objective.

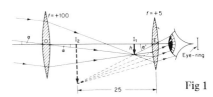

Fig 1

Fig. 1. Suppose the distance of I_1 from the eyepiece E = u. Then, since $v = -25$ and $f = +5$ using the 'real is positive' convention,

$$\frac{1}{(-25)} + \frac{1}{u} = \frac{1}{(+5)},$$

from which

$$u = \frac{25}{6} = 4\tfrac{1}{6} \text{ cm} = I_1 E$$

The angular magnification, M, = $\dfrac{\alpha'}{\alpha}$.

Now $\alpha' = h/I_1 E$, where h is the length of the image at I_1, and α = angle subtended by distant object at $O = h/OI_1$

$$\therefore M = \frac{h/I_1 E}{h/OI_1} = \frac{OI_1}{I_1 E} = \frac{100}{4\frac{1}{6}} = 24.$$

(ii) The best position of the eye is at the *eye-ring,* where it will collect all the rays passing through the objective and eyepiece. The eye-ring is the image of the objective lens in the eyepiece. Thus the eye-ring distance from E, v, is given by

$$\frac{1}{v} + \frac{1}{(+104\frac{1}{6})} = \frac{1}{(+5)},$$

from which $v = 5.3$ cm.

3 (i) A converging meniscus lens has radii of curvature 5 and 6 cm respectively and the glass has a refractive index 1.5. Calculate the focal length of the lens.

(ii) An equiconvex lens has a focal length of 10 cm and is made of glass of refractive index 1.6. Calculate the radius of curvature of the surfaces. If the diameter of the lens is 4 cm, what is the least thickness of the lens at its centre?

(i) The surface of radius 6 cm is concave to the air and that of 5 cm is convex to the air.

(i) *Real is positive.* $r_1 = +5$cm, $r_2 = -6$ cm.

$$\therefore \frac{1}{f} = (n-1)\left(\frac{1}{r_1} + \frac{1}{r_2}\right) = (1.5-1)\left[\frac{1}{(+5)} + \frac{1}{(-6)}\right]$$

$$= (1.5-1)(\tfrac{1}{5} - \tfrac{1}{6}) = \tfrac{1}{60}$$

$$\therefore f = +60 \text{ cm}.$$

(ii) If r is the numerical value of the radius of each surface, then

$$\frac{1}{f} = (n-1)\left(\frac{1}{r} + \frac{1}{r}\right) = (n-1)\frac{2}{r}$$

$$\therefore r = 2(n-1)f = 2(1.6-1) \times 10 = 12 \text{ cm}.$$

The height y of the lens from the centre = $\frac{1}{2} \times 4$ cm = 2 cm. If h is *half* the least thickness at the centre, then, from the product of chords of a circle,

$$h(2r - h) = y^2,$$

or if h is small compared with $2r$,

$$h = \frac{y^2}{2r} = \frac{2^2}{2 \times 12} = \tfrac{1}{6} \text{ cm} = 1.7 \text{ mm (approx.)}$$

$$\therefore \text{least thickness} = 2h = 3.4 \text{ mm}$$

4 A 60° glass prism of $n = 1.50$ is placed in water of $n = 1.33$. Calculate the angle of minimum deviation for light refracted through the prism.

The prism is now placed in air. Find the angle of incidence of a ray in air when the angle of deviation is a maximum. Find also the greatest angle of the prism for light to be refracted through the prism.

First Part
$$n = \frac{\sin\frac{A+D}{2}}{\sin\frac{A}{2}},$$

where n is the *relative* refractive index between glass and water. Since $n = 1.5/1.33$,

$$\therefore \frac{1.5}{1.33} = \frac{\sin\left(\frac{60+D}{2}\right)^\circ}{\sin\frac{60^\circ}{2}}$$

$$\therefore \sin\left(\frac{60+D}{2}\right)^\circ = \frac{1.5}{1.33}\sin 30^\circ = 0.5639$$

$$\therefore \frac{60^\circ + D}{2} = 34^\circ\,21'$$

$$\therefore D = 8^\circ\,42'.$$

Second Part. The angle of deviation is a maximum when the ray emerges at B along the prism surface (grazing emergence). Fig. 2 (i). Thus the angle of incidence at B in the glass is the critical angle, c. Now

$$\sin c = \frac{1}{n} = \frac{1}{1.5} = 0.6667$$

$$\therefore c = 41^\circ 49'.$$

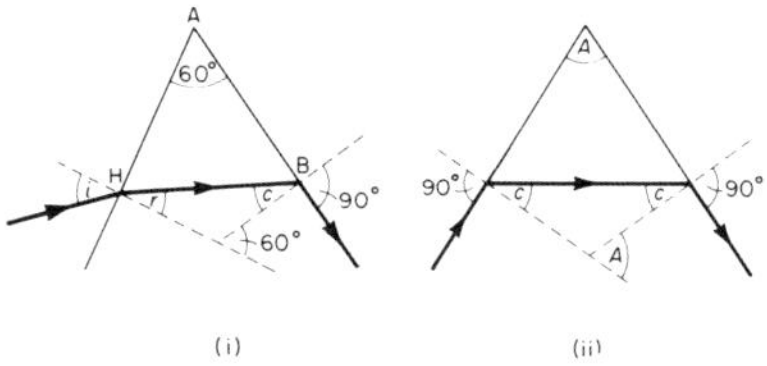

Fig. 2

The angle of refraction r in the glass at H $= 60^\circ - c = 18^\circ\,11'$. Since $\sin i = n \sin r$.

$$\therefore \sin i = 1.5 \sin 18^\circ\,11' = 0.4682$$

$$\therefore i = 27^\circ\,55'.$$

Third Part. The greatest (limiting angle) angle, A, of the prism corresponds to the case of grazing incidence and grazing emergence. Fig. 2 (ii).

$$\therefore A = c + c = 2c,$$

where c is the critical angle.

$$\therefore A = 2 \times 41^\circ\,49' = 83^\circ\,38'.$$

PHYSICAL OPTICS

5 A diverging lens of focal length 12 cm is placed 20 cm to the right of a converging lens of focal length 10 cm. An object is now placed 15 cm to the left of the converging lens. Find the final image formed by refraction through both lenses, and the magnification.

For the *converging lens*, $f = +10$, $u = +15$, using the 'real is positive' convention. The image distance, v, is given by

$$\frac{1}{v} + \frac{1}{(+15)} = \frac{1}{(+10)}$$

from which $v = +30$ cm

For the *diverging lens*, $f = -12$, $u = -(30 - 20) = -10$, since the image is a virtual object for this lens. The image distance, v, is given by

$$\frac{1}{v} + \frac{1}{(-10)} = \frac{1}{(-12)},$$

from which $v = 60$ cm.

MAGNIFICATION. Since $m = \frac{v}{u}$, the magnification produced by the converging lens $= 30/15 = 2$ and that produced by the diverging lens $= 60/10 = 6$.

∴ total magnification $= 2 \times 6 = 12$.

6 A diffraction grating with 500 lines per millimetre is illuminated normally by yellow light of wavelength 5.89×10^{-7} m. (i) At what angle to the normal is the first order image? (ii) What is the highest order in which an image may be obtained? (iii) If the grating is also illuminated normally by light of wavelength 4.5×10^{-7} m, what is the separation of the two first order images in the plane of the objective of a telescope of 20 cm focal length used to view them.

(i) Grating spacing, d, $= \frac{1}{500}$ mm $= \frac{1}{500 \times 10^3} = 2 \times 10^{-6}$ m

First order image is given by $d \sin \theta = \lambda$

$$\therefore \sin \theta = \frac{\lambda}{d} = \frac{5.89 \times 10^{-7}}{2 \times 10^{-6}} = 0.294$$

$$\therefore \theta = 17° 8'$$

(ii) Highest order m is given by $d \sin \theta = m\lambda$, or

$$m = \frac{d \sin \theta}{\lambda}$$

Largest value of $\sin \theta = 1$, when $\theta = 90°$. In this case

$$m = \frac{d}{\lambda} = \frac{2 \times 10^{-6}}{5.89 \times 10^{-7}} = 3.4$$

Thus highest order for $m = 3$.

74 GRADED EXERCISES AND WORKED EXAMPLES IN PHYSICS

(iii) For $\lambda = 4.5 \times 10^{-7}$ m, angular position θ of first order image is given by

$$d \sin \theta = \lambda,$$

or

$$\sin \theta = \frac{\lambda}{d} = \frac{4.5 \times 10^{-7}}{2 \times 10^{-6}} = 0.225$$

$$\therefore \theta = 13° 0'$$

Hence, from (i), angular separation of two wavelengths = $17° 6' - 13° 0'$ = $4° 6'$.

$\therefore$ length between images = $f \tan 4° 6'$

$$= 20 \tan 4° 6' = 1.4 \text{ cm} = 14 \text{ mm}$$

7 What do you know about the speeds of red and blue light (i) *in vacuo*, (ii) in glass?

Calculate the dispersion between the red and blue rays when white light is incident at a small angle on a crown-glass prism of angle $8°$. Find also the angle of a flint-glass prism which makes a direct-vision spectroscope with this crown-glass prism. [$n_b = 1.532$, $n_r = 1.518$, n (yellow) = 1.524 for crown glass; $n_b = 1.665$, $n_r = 1.645$, n (yellow) = 1.652 for flint glass.]

First Part. The sun appears white to an observer on the earth. Thus the speeds of red and blue light in empty space (as well as other colours) are the same. The refractive index of glass = velocity of light *in vacuo* $\div$ velocity in glass. Since the refractive index of glass for blue light is greater than for red light, the speed of blue light in glass must be less than that of red light.

Second Part. The deviation of light by a prism of small angle A is given by $d = (n - 1)A$.

dispersion = angle between emerging red and blue rays

$$= d_b - d_r = (n_b - 1)A - (n_r - 1)A$$

$$= (n_b - n_r)A = (1.532 - 1.518)8° = 0.112°.$$

For a direct-vision spectroscope dispersion is necessary, but there must be *no deviation* of the incident beam. To achieve this, a crown-glass and a flint-glass prism are joined together with their refracting angles pointing in opposite directions; the deviation of the average light of the beam, chosen as the yellow light, must then be equal for both prisms. Thus if A is the flint-glass prism angle,

$$(1.524 - 1)8° = (1.652 - 1)A$$

$$\therefore A = 6.4°.$$

6
Waves. Sound

32. WAVES

1 Define *wavelength*. The wavelength of a progressive wave is 0.3 m. What is the phase difference between two points (i) 0.1 m apart, (ii) 0.18 m apart, (iii) 1 m apart?

2 A water wave travelling in a constant direction has a wavelength of 12 mm, a speed of 0.24 m s^{-1} and an amplitude of 5 mm. Calculate the frequency of the wave and the maximum velocity of a particle of water.

3 The frequency of a wave is 600 Hz and velocity is 340 m s^{-1}. Find the phase difference between two points (i) 0.17 m apart, (ii) 1.7 m apart.

4 Define *frequency* and *velocity* of a wave. Derive the relation between them.

5 A plane progressive wave in the x-direction is represented by $y = a \sin 2\pi(vt - x)/\lambda$. What are y, a, λ and v? What is the *period* of the wave-motion in terms of two of these four quantities?

Write down the equation of a wave travelling in the negative x-direction with the same velocity wavelength and amplitude.

6 *Sound waves are matter waves and are longitudinal. Light waves are electromagnetic waves and are transverse.*

Explain the meaning of these statements, drawing sketches to illustrate your answer.

7 A plane progressive wave in the x-direction is represented by $y = a \sin(\omega t - \beta x)$. Write down expressions for the velocity v, wavelength λ and period T of the wave in terms of ω and β.

8 A plane progressive wave in the x-direction is expressed by $y = a \sin(2000\pi t - 100\pi x/17)$. Calculate the frequency, wavelength and velocity of the wave is y and x are in metres and t in seconds.

9 A progressive wave has a velocity of 10 000 m s^{-1}, a frequency of 20 000 Hz and an amplitude of 2 mm. Obtain an expression for the wave.

10 What is the difference between a *longitudinal* and a *transverse* wave? Identify the following waves (i)–(v) from the data given:

(i) Transverse—wavelength 6×10^{-7} m
(ii) Longitudinal—wavelength 1 m
(iii) Transverse—wavelength 50 mm
(iv) Longitudinal—wavelength 10^{-3} m
(v) Transverse—wavelength 200 m.

Choose your answer from one of the following letters **A–F**:
A radio waves,
B light waves,
C sound waves in air,
D ultrasonic waves in water,
E X-rays,
F water waves in ripple tank.

11 Show by dimensions that the velocity of a transverse wave along a string is given by $v = k\sqrt{T/m}$, where T is the tension, m is the mass per unit length and k is a constant. Assuming $k = 1$, calculate the velocity of the wave when the tension is 50 N, the length of the string is 1.5 m and its mass is 3×10^{-4} kg.

12 Show by dimensions that the velocity of a longitudinal wave in a medium is given by $v \propto \sqrt{E/\rho}$, where E is the modulus of elasticity and ρ the density of the medium. Water has a bulk modulus 2×10^9 N m^{-2} and a density of 1000 kg m^{-3}. Assuming $v = \sqrt{E/\rho}$, calculate the time taken for a wave in water to travel 50 metres.

13 Define *node* and *antinode* of a stationary wave. If the distance between a node and the nearest antinode is 15 cm, calculate the frequency of the wave. (Velocity of wave = 330 m s^{-1}.)

14 The distance between successive antinodes in a stationary wave is 20 cm. Find the velocity of the wave if the frequency is 800 Hz.

15 Describe fully how a stationary wave differs from a plane-progressive wave. What do you know about the condition of the air between a whistle and a smooth wall when the whistle is sounding?

16 With the help of diagrams, or otherwise, explain how two plane-progressive waves may produce a stationary wave.

17 A stationary wave in a direction Ox is represented by $y = B \sin 200\, \pi t$, where y is the displacement in metre, t is the time in second, and B is an *amplitude which varies with distance* x according to the relation $B = 10^{-6} \cos 2x$, where B is in metre. Find (i) the frequency, (ii) the wavelength, (iii) the amplitude at the antinodes, (iv) the distance between two consecutive nodes, (v) the phase difference between two points (a) 1 m apart, (b) 2 m apart.

18 Simple pendula of varying lengths are suspended from a horizontal thread. Describe and explain how this system can be used to demonstrate *forced* and *resonant* oscillations.

19 What is the difference between a *damped* oscillation and a *resonant* oscillation. Draw sketches to illustrate your answer.

Give one example each of mechanical resonance, electrical resonance and optical resonance.

33. SOUND

1 Define the following terms used in connection with a sound wave: *frequency, amplitude, wavelength, compression, rarefaction, particle displacement*.

2 Describe the main features of a *plane-progressive sound wave*. In your answer, draw sketches showing how (i) the displacement, (ii) the pressure and (iii) the particle velocities vary along the medium as the wave travels.

3 On what factors do the *pitch, loudness* and *timbre* of a note depend. Draw diagrams to illustrate your answer.

4 Describe an experiment to illustrate (i) the reflection of sound, (ii) the refraction of sound, (iii) the interference of two sound waves.

5 Explain why sounds are heard more easily (i) when the wind is blowing towards the observer, (ii) over long distances on a clear night. Draw diagrams of the wavefronts in your answer.

6 Two loudspeakers, each emitting a note of frequency 500 Hz, are placed facing each other on opposite sides of a room. Describe and explain what is heard at points in the air between the two speakers. (Velocity of sound = 340 m s^{-1})

7 Two simple harmonic notes A and B produce a *Lissajous figure* on an oscilloscope screen by using two microphones connected to the X- and Y-plates respectively.
What do you know about A and B when the Lissajous figure is (i) a straight inclined line, (ii) a figure of 8, (iii) an ellipse?

8 A small source of sound produces a sound intensity 5×10^{-3} W m^{-2} at a place 10 m away. Calculate the power of the source in watt and state your assumptions. What is the intensity at a distance of 20 m away?

9 The amplitude of vibration of air at a distance of 8 m from a small source of sound is 0.1 mm. What is the amplitude at a distance of 4 m? If the source has a power of 50 milliwatts, calculate the intensity at a distance of 10 m away.

Velocity of Sound

10 The wavelength of a note is (i) 0.4 m, (ii) 1.0 m. Calculate the frequency in each case. (Velocity of sound = 340 m s^{-1}.)

11 The velocity of sound in air at 20°C is 344 m s^{-1}. Calculate the velocity (i) at 15°C, (ii) at 0°C. What is the velocity at 20°C if the barometric pressure changes from 760 to 750 mmHg?

12 Find the wavelength due to a note of frequency 300 Hz in air at a temperature of 12°C. (Velocity of sound in air at 0°C = 331 m s^{-1}.)

13 Newton's formula for the velocity of sound in air is $V = \sqrt{p/\rho}$, where p is the air pressure and ρ is the air density. Though incorrect, show that the formula is *dimensionally* correct.

Using the correct formula, calculate the velocity of sound in air if the density is 1.29 kg m^{-3} at s.t.p. and the ratio of the molar heat capacities of air is 1.4.

14 Prove that the velocity of sound in a gas is proportional to the square root of the absolute temperature and is independent of the pressure.

15 A whistle of frequency 500 Hz is blown at one end of a closed corridor, and the fifth echo is heard 2 seconds after. If the wavelength of the sound is 0.66 m, calculate the length of the corridor.

Beats

16 What are 'beats'? Two tuning-forks of frequency 256, 260 Hz respectively are sounding in air. Find the number of beats per second produced. Draw sketches of the wave form of the two notes and of their resultant.

17 Two tuning-forks, A, B, produce three beats per second when sounding. If A has a frequency of 512 Hz what are the possible frequencies of B? Explain how you would determine the actual frequency of B.

18 Explain how beats are produced, illustrating your answer with wave diagrams. Prove that the beat frequency is the difference in frequency of the two vibrations concerned.

Doppler's Principle

19 A train sounding a whistle of 800 Hz approaches a stationary observer at 30 m s^{-1}. If the velocity of sound in air is 340 m s^{-1}, calculate from first principles the apparent frequency of the whistle to the observer. What is the apparent frequency after the train passes the observer?

20 A man, X, in a car drives away from a stationary observer Y with a velocity of 15 m s^{-1}, blowing a horn whose note has a frequency of 1000 Hz. Calculate from first principles the apparent frequency of the note heard by Y.

What is the apparent frequency if X approaches Y with a velocity of 30 m s^{-1} while sounding the horn? (Velocity of sound = 340 m s^{-1}.)

21 A car with a velocity of 30 m s^{-1} approaches a train which is sounding a whistle of 900 Hz. If the train is (i) stationary, (ii) moving with a velocity of 15 m s^{-1} in the same direction as the car, calculate the frequency of the note heard in each case by a man in the car, and explain the principles of your calculation. (Velocity of sound = 340 m s^{-1}.)

22 Two observers, A, B, have each sources of sound of frequency 1000 Hz. Calculate the beats heard by B if A moves away from him with a velocity of 2 m s^{-1}. What are the beats heard by A? (Velocity of sound = 340 m s^{-1}.)

23 A person carrying a whistle sounding at 600 Hz moves towards a wall with a velocity of 1 m s^{-1}. If the wall is smooth and perpendicular to the direction of motion of the whistle, calculate the beats heard by the person. (Velocity of sound = 340 m s^{-1}.)

24 A stationary observer is approached by a source of sound, and the pitch of the note is increased in the ratio 15:14. Calculate the velocity of the source if the velocity of sound is 360 m s^{-1}. What is the new velocity if the pitch is decreased in the ratio 14:15 when the source recedes from the observer?

34. WAVES IN PIPES, STRINGS, RODS

Pipes

1 A closed organ pipe has a length of 0.6 m. Calculate its fundamental frequency and that of its first overtone. Describe in each case the vibration of the air in the pipe, drawing diagrams in illustration. (Velocity of sound = 336 m s^{-1}.)

2 An open organ pipe has a length of 0.5 m. Calculate the fundamental frequency and that of its first overtone. Describe in each case the vibration of the air in the pipe, drawing diagrams in illustration. (Velocity of sound = 340 m s^{-1}.)

3 When a sounding tuning-fork of frequency 500 Hz is placed over a column of water in a pipe, resonance is obtained at lengths of 15.2 and 46.8 cm respectively of the air-column. Find the velocity of sound in air and the end-correction of the pipe.

4 Calculate the frequency of the first overtone in an open pipe 2 m long having an end-correction of 1.5 cm, if the temperature of the air is 27°C. (Velocity of sound at 0°C = 332 m s^{-1}.)

5 Describe the pressure variation in a closed pipe when it is sounding (i) its fundamental note, (ii) its first overtone. How has the presence of nodes and antinodes in pipes, and the pressure variation, been demonstrated?

6 How is the pitch of a note from a pipe affected by (i) its length, (ii) its diameter, (iii) the temperature of the air in the pipe? Give reasons for your answers.

7 Two open pipes, 1.00 m and 1.02 m long respectively, produce 3.3 beats per second when they are sounding their fundamental notes. Neglecting the end-corrections, calculate the velocity of sound in air and the frequencies of the fundamentals.

8 Describe how the velocity of sound in air can be determined by a

resonance tube experiment (i) if only one tuning-fork is available, (ii) if six tuning-forks of varying frequency are available and only one resonance position is allowed.

Strings

9 The length of a stretched string is 0.4 m; its mass is 1.0×10^{-4} kg. If the tension in the string is 10 N, calculate (i) the velocity of the transverse wave in the string, (ii) the frequency of its fundamental note, (iii) the frequency of the first overtone when plucked in the middle.

10 The note produced by a wire 0.5 m long is four times the frequency of a note produced by another wire 1.5 m long when each is plucked in the middle. Find the relative mass per unit length of the two wires if the tension is the same.

11 Two identical wires in a sonometer have each a tension of 60 N and a frequency of 300 Hz when plucked in the middle. Calculate the beat frequency if the tension in one wire is increased by 2 N and both wires are plucked in the middle.

12 A wire has a tension of 20 N, and produces a note of 100 Hz when plucked in the middle. Find its diameter if it is 1 m long and density 6000 kg m^{-3}. State the frequency of the first overtone.

13 Assuming the formula for the velocity of a transverse wave in a stretched string, derive the expression for the fundamental frequency of the string. Describe briefly how you would verify the relation between (i) the frequency and length, (ii) the frequency and tension in a sonometer wire.

14 A wire 0.75 m long is touched lightly with a feather one-third from one end, and is bowed near that end. Calculate the tension if the note obtained has a frequency of 200 Hz and the mass per metre is 2×10^{-5} kg m^{-1}.

15 A sonometer wire has a length of 1.12 m and tension of 50 N. When a.c. mains current is passed into the wire, the wire is set into resonance in its fundamental mode by a perpendicular magnetic field at its midpoint, after adjustment of a bridge. The wire has a mass per unit length of 4×10^{-3} kg m^{-1}. Find the frequency of the a.c. mains.

16 A stretched wire 1 metre long is (i) plucked transversely in the middle, (ii) stroked longitudinally. If the frequency is 500 Hz in the first case and 3000 Hz in the second case, calculate the velocity of transverse and longitudinal waves in the wire. Describe the vibration of the wire in each case.

Rods

17 In a dust (Kundt) tube apparatus, a brass rod 0.80 m long is fixed at the middle and stroked longitudinally. The heaps of powder in the air in the tube are 0.04 m apart on the average at resonance, and the air temperature is 17°C. Calculate the velocity of sound in brass, and the Young modulus for brass. (Velocity of sound in air at 0°C = 332 m s^{-1}, density of brass = 8500 kg m^{-3}.)

18 In a dust tube apparatus, a glass rod 1 metre long is fixed at the middle and stroked longitudinally. If the heaps of powder in the air in the tube are then 3 cm apart on the average, find the velocity of sound in glass. When the air in the tube is replaced by carbon dioxide gas, the heaps are then 4 cm apart. Find the velocity of sound in carbon dioxide gas. (Velocity of sound in air = 340 m s^{-1}.)

19 Describe how the ratio of the molar heat capacities of a gas can be measured by a dust tube experiment. Give the theory.

20 In the dust tube experiment, the heaps of powder occupy a total distance of 0.90 m and are 0.10 m apart when the air in the tube is set into resonance at a frequency f. Calculate f if the velocity of sound is 340 m s^{-1}.

Find also (i) the number of wavelengths in the air inside the tube, (ii) the next higher frequency when the air in the tube is again set into resonance.

35. MULTIPLE CHOICE QUESTIONS – WAVES, SOUND

*For each question, choose one statement from **A** to **E** which is the most appropriate.*

1 If λ is the wavelength of a wave and T is the period, the speed of the wave is given by
A λT, **B** λ/T, **C** λT^2, **D** T/λ, **E** λT^4.

2 The pressure and temperature at the top X of a mountain is 720 mmHg and 7°C respectively, and at the foot Y it is 760 mmHg and 17°C respectively. The ratio of the velocity of sound at X and Y is
A $\sqrt{28/29}$, **B** 720/760, **C** $\sqrt{7/17}$, **D** 720 × 7/760 × 17, **E** 720 × 280/760 × 290.

3 The tension in a wire is 80 N, and 10 m of the wire has a mass of 0.02 kg. The speed of a transverse wave in the wire is
A 160 m s^{-1}, **B** 40 m s^{-1}, **C** 40 000 m s^{-1}, **D** 80 m s^{-1}, **E** 200 m s^{-1}.

4 Two similar vibrating strings will produce beats if
A the amplitudes of vibration differ slightly,
B the lengths differ slightly,
C the frequencies are equal,
D the wavelengths of the sound produced are appreciably different,
E the waveforms of the sound are different.

5 A plane-progressive sound wave is represented by the equation $y = 10^{-4} \sin 2\pi(100t - x/3.4)$, where y, x are in metre and t in second. In this case
A the frequency is 100 Hz and the wavelength about 0.3 m,

B = the amplitude of vibration is 10^{-2} m,
C the velocity of the wave is 340 m s^{-1},
D the frequency is 200π Hz,
E the wavelength is $2\pi/3.4$ m.

6 A closed pipe has a length of 0.34 m and negligible end-correction, and is sounding its fundamental. The velocity of sound is 340 m s^{-1}. Then
A the pressure node is at the closed end,
B the wavelength of the wave is given by $\lambda/2 = 0.34$ m,
C the displacement antinode is at the closed end,
D the frequency of the note is 250 Hz,
E a plane-progressive wave is produced inside the pipe.

7 A train sounding a whistle of 2000 Hz moves towards a stationary observer with a velocity of 20 m s^{-1}. The speed of sound is 340 m s^{-1}. The observer then hears a sound of frequency
A 6800 Hz, **B** 1880 Hz, **C** 34 000 Hz, **D** 2125 Hz, **E** 2360 Hz.

8 The following properties of a sound wave may be altered when circumstances are changed, as in (a), (b), (c), (d), (e), (f) below:
(i) frequency, (ii) velocity, (iii) loudness.
What is the appropriate answer then,
(a) A tuning fork sounds and the temperature of the air increases
(b) A tuning fork sounds and the pressure of the air decreases
(c) An organ pipe sounds and the temperature of the air increases
(d) A tuning fork sounds and humidity increases
(e) An organ pipe sounds and humidity increases
(f) Sound made by a vibrating string, length altered.
Answer
A if (i) only is changed,
B if (ii) only is changed,
C if (iii) only is changed,
D if (i) and (ii) only are changed and
E if (ii) and (iii) only are changed.

WORKED EXAMPLES ON WAVES AND SOUND

1 The displacement y of a plane-progressive wave is given by

$$y = 10^{-4} \sin(200\pi t - 0.5\pi x),$$

where y and x are in metre and t in second. Find (i) the amplitude, wavelength and velocity of the wave, (ii) the phase difference between two points 1 m apart.

(i) Amplitude = maximum value of $y = 10^{-4}$ m.
Comparing the equation with $y = a \sin 2\pi(vt - x)/\lambda$, then

$$\frac{v}{\lambda} = 100 \quad \text{and} \quad \frac{2}{\lambda} = 0.5$$

$$\therefore \lambda = \frac{2}{0.5} = 4 \text{ m}$$

and
$$v = 100\lambda = 400 \text{ m s}^{-1}.$$

(ii) Points λ apart have a phase difference 2π. Since $\lambda = 4$ m, two points 1 metre apart have phase difference $\frac{1}{4} \times 2\pi = \frac{\pi}{2}$.

2 Calculate the frequency of the second overtone of a pipe, 0.80 m long, closed at one end when the air temperature is 17°C. The end-correction is 0.015 m and the velocity of sound in air at 0°C is 331 m s^{-1}.

The frequencies of the notes obtainable from a closed pipe are f_0, $3f_0$, $5f_0$, $7f_0$, and so on, where f_0 is the fundamental frequency. In the case of the second overtone (a frequency of $5f_0$) the pipe contains $1\frac{1}{4}$ wavelengths.

$$\therefore \quad 1\tfrac{1}{4}\lambda = l + c = 0.80 + 0.015 = 0.815 \text{ m}$$

$$\therefore \quad \lambda = \frac{0.815 \times 4}{5} = 0.652 \text{ m}$$

The velocity in air is proportional to $\sqrt{T}$.

$$\therefore \quad \text{velocity, } V, \text{ at } 17°\text{C} = 331\sqrt{\frac{273 + 17}{273}}$$

$$= 331\sqrt{\frac{290}{273}} \text{ m s}^{-1}$$

$$\therefore \quad \text{frequency } f, = \frac{V}{\lambda} = \frac{331}{0.652}\sqrt{\frac{290}{273}}$$

$$= 523 \text{ Hz}.$$

3 The wavelength in air of a note of frequency 800 Hz was found by experiment to be 0.425 m, the temperature being 15°C. Calculate the ratio of the molar heat capacities of air if the density of air at s.t.p. is 1.29 kg m^{-3}.

The velocity V_1 at 15°C = $f\lambda$ = 800 × 0.425 m s^{-1} = 340 m s^{-1}. Since velocity $\propto$ square root of absolute temperature,

$$\frac{V_0}{V_1} = \sqrt{\frac{273}{273 + 15}},$$

where V_0 is the velocity at 0°C.

$$\therefore \quad V_0 = 340\sqrt{\frac{273}{288}} = \sqrt{\frac{\gamma p}{\rho}}$$

$$\therefore \quad \gamma = \frac{340^2 \times 273 \times \rho}{288 \times p}.$$

Now
$$p = 0.76 \times 13\,600 \times 9.8 \text{ N m}^{-2}$$

and
$$\rho = 1.29 \text{ kg m}^{-3}$$
$$\therefore \gamma = \frac{340^2 \times 273 \times 1.29}{288 \times 0.76 \times 13\,600 \times 9.8} = 1.4$$

4 What is *Doppler's Principle*? An observer moving at 15 m s^{-1} approaches a stationary source of sound of frequency 1000 Hz. Calculate the apparent frequency of the source. What is the apparent frequency if the observer is stationary and the source moves away from him with a velocity of 54 km per hour? (Velocity of sound = 340 m s^{-1}.)

First Part. Doppler's Principle states that the frequency of a source of light or sound is apparently altered whenever there is a relative velocity between the source and the observer.

Second Part. (i) Since frequency = velocity of wave/wavelength, the apparent frequency f' is given by

$$f' = \frac{V'}{\lambda'},$$

where V' is the velocity of sound relative to the observer and λ' is the wavelength of the waves reaching the observer.

Now
$$V' = 340 + 15 = 355 \text{ m s}^{-1},$$
since the observer moves towards the sound waves with a velocity of 15 m s^{-1}. Also, the source is stationary, so that the wavelength λ of the waves reaching the observer = velocity of sound/frequency, or

$$\lambda' = \frac{340}{1000}$$

$$\therefore f' = \frac{V'}{\lambda'} = \frac{355}{340/1000} = \frac{355 \times 1000}{340} = 1044 \text{ Hz}.$$

(ii) When the source is stationary, $V' = 340$. The wavelength λ' of the waves reaching the observer is given by $\lambda' = (340 + 15)/1000$, since the source moves a distance of 15 m in 1 second away from the observer.

$$\therefore \text{apparent frequency}, f' = \frac{V'}{\lambda'} = \frac{340}{355/1000}$$

$$= \frac{340 \times 1000}{355} = 958 \text{ Hz}.$$

5 A stretched wire is damped lightly one-quarter of its length from one end, and plucked near that end. The frequency of the note is 600 Hz. Calculate the length of the wire if the tension is 80 N and the mass per metre is 2 × 10^{-3} kg. (Assume $g = 10$ m s^{-2}.)

WAVES. SOUND

The distance between consecutive nodes, N, of the wave is $l/4$, where l is the length of the wire.

$$\therefore NN = \frac{\lambda}{2} = \frac{l}{4}, \text{ or } \lambda = \frac{l}{2}$$

$$\therefore f = \frac{V}{\lambda} = \frac{\sqrt{\frac{T}{m}}}{l/2} = \frac{2}{l}\sqrt{\frac{T}{m}}$$

Now $T = 80$ N, $m = 2 \times 10^{-3}$ kg m^{-1}, $f = 600$ Hz.

$$\therefore l = \frac{2}{f}\sqrt{\frac{T}{m}}$$

$$= \frac{2}{600}\sqrt{\frac{80}{2 \times 10^{-3}}} = \frac{2}{3} \text{ m}$$

6 How is the pitch of a note from an organ pipe affected by (i) the temperature of the air, (ii) the diameter of the pipe? Two open organ pipes 1.00 m and 1.01 m long give 17 beats in 10 seconds when each is sounding its fundamental. Calculate a value for the velocity of sound and the frequencies of the fundamentals.

First Part. The frequency, $f, = \frac{V}{\lambda}$, where V is the velocity of sound in air and λ is the wavelength. (i) V increases as the temperature increases; hence the pitch increases as the temperature increases. (ii) The wavelength increases as the diameter increases, since the end-correction = 0.6 × radius. Thus the pitch decreases as the diameter increases.

Second Part. The length of an organ pipe sounding its fundamental is $\lambda/2$ (the distance between consecutive antinodes). Hence $\lambda = 2l = 2.00$ m.

$$\therefore \text{ frequency, } f_1, \text{ of first note} = \frac{V}{\lambda} = \frac{V}{2.00}$$

and frequency, f_2, of second note = $V/2.02$

$$\therefore \frac{V}{2.00} - \frac{V}{2.02} = \text{beat frequency} = \frac{17}{10}$$

$$\therefore V = \frac{2.00 \times 2.02 \times 17}{0.02 \times 10} = 343.4 \text{ m s}^{-1}$$

Also,

$$f_1 = \frac{V}{2.0} = \frac{343.4}{2.0} = 171.7 \text{ Hz}$$

and

$$f_2 = \frac{V}{2.02} = \frac{343.4}{2.02} = 170 \text{ Hz}.$$

7
Electricity

36. ELECTROSTATICS—FORCE AND POTENTIAL
(Assume $\epsilon_0 = 8.85 \times 10^{-12}$ F m^{-1}, or $1/4\pi\epsilon_0 = 9 \times 10^9$ m F^{-1} approximately.)

1 Calculate the force between charges of $+10^{-9}$ C and -4×10^{-9} C when they are 2×10^{-2} m apart in air. What is the distance between the charges when the force between them is 10^{-5} N?

2 Find (i) the electric intensity at a point 10^{-3} m from a positive charge of 2×10^{-8} C, (ii) the magnitude and direction of the force on a negative charge of 8×10^{-19} C if placed at that point.

3 Draw sketches showing the lines of electric force (i) between two positive charges, (ii) round an isolated negative charge, (iii) between a positive charge and an earthed plate, (iv) between two parallel plates having equal and opposite charges.
In your sketches, indicate by X where the electric field may be *uniform*.

4 Draw a labelled diagram of a *leaf electroscope*. Why is the leaf surrounded by a metal can? Describe with the aid of diagrams how an electroscope is charged negatively by induction and explain what happens.

5 How would you compare the charges on two metal spheres by means of a leaf electroscope? Describe experiments which demonstrate (i) electrostatic screening, (ii) that equal and opposite charges are produced when a polythene rod is rubbed with a duster.

6 Describe Faraday's ice-pail experiment. What conclusions can be drawn from the experimental results?

7 At a point 1 cm (10^{-2} m) from a positive point charge Q, the intensity of the electric field is 14,400 V m^{-1} and the potential is 144 V.
(i) Find the magnitude of Q. (ii) Calculate the electric intensity E and the potential V at points distant r = 2, 3, 4, 5, 6, 8, 10 cm respectively from the charge. From your values, plot graphs of E against r and of V against r, and comment on the shapes of the two graphs.

8 Two equal and opposite point charges of 400 μC are 3 mm apart. Calculate (i) the electric intensity, (ii) the electric potential at a point 5 mm from the positive charge and 4 mm from the negative charge.

9 A sphere A of radius 5×10^{-2} m has a charge of $+10 \times 10^{-9}$ C; a sphere B of radius 4×10^{-2} m has a charge of $+12 \times 10^{-9}$ C. Calculate the potential difference between A, B.

86

Which way will electrons flow between A and B if they are joined by a wire?

10 Two points X and Y in a uniform electric field are 5 mm apart. When a charge of 10^{-9} C is moved between X and Y, 10^{-6} J of work is done. Calculate (i) the p.d. between X and Y, (ii) the electric intensity in the field, (iii) the force on the charge in the field.

11 Two parallel plates, 20 mm apart, are connected to a 1000 volt battery. Calculate (i) the electric intensity between the plates (ii) the speed acquired by an electron in moving through the field from one plate to another in empty space if its initial speed is zero. (Mass of electron = 9.0×10^{-31} kg; charge on electron = 1.6×10^{-19} C.)

12 The p.d. between the cathode K and anode A of an 'electron gun' is 5000 V. Calculate the energy gained by 10^6 electrons which move from K to A, if the charge on an electron is 1.6×10^{-19} C.

Assuming an electron starts from zero velocity at K, find its velocity on reaching A if the mass of an electron is 9×10^{-31} kg.

13 A p.d. of 150 V is connected between the cathode and anode of an electron tube. An electron of mass 9×10^{-31} kg and charge 1.6×10^{-19} C moves from cathode to anode, starting with zero velocity. Calculate the velocity of the electron when it just reaches the anode.

14 Describe an *electrostatic generator*, such as the Van de Graaff generator. Explain (i) how the 'action at points' is utilized, (ii) the source of energy of the charge stored on the high voltage terminal X, (iii) why there is no repulsive force on charges which move towards X from the pointed conductors.

15 A van de Graaff generator reaches a maximum potential of 36 000 V. Calculate the charge on its sphere if its radius is 0.2 m.

To maintain this potential against leakage of charge through the insulating pillar, a charge of 10^{-3} C must be added every second to the sphere. Calculate the ohmic resistance of the pillar.

16 An oil drop carries a charge 4*e*, where *e* is the electronic charge, 1.6×10^{-19} C. If the drop is stationary between two plates 5 mm apart and having a p.d. of 600 V, calculate the mass of the drop. (Assume $g = 10$ m s^{-2}.)

17 An electron, charge 1.6×10^{-19} C, moves from rest in empty space between two points 3 mm apart whose p.d. is 150 V. Find (i) the electric intensity of the field, (ii) the energy acquired by the electron, (iii) the final velocity of the electron if its mass is 9×10^{-31} kg.

18 Define *potential gradient, electric intensity*. Derive the relation between them.

What are *equipotentials* in an electric field? Explain why the equipotentials are perpendicular to the electric lines of force where they cross each other.

19 Define *surface-density* of charge. Draw sketches of the variation of surface-density round (i) a charged metal sphere, (ii) a pear-shaped conductor. How would you measure the variation of surface-density of a pear-shaped conductor?

20 A hollow metal sphere of radius 5 cm has a charge of 6×10^{-9} C. Calculate (i) the surface density, (ii) the intensity at a point inside the sphere, (iii) the potential inside the sphere, (iv) the intensity just outside the sphere, (v) the force per unit area on the sphere.

21 Three concentric spheres, A, B, C, have radii of 5, 10, 20 cm respectively, and charges of $+2$, -1 and $+4 \times 10^{-8}$ C respectively. Calculate the respective potentials of A, B, C.

37. CAPACITORS

(Assume $\epsilon_0 = 8.85 \times 10^{-12}$ F m^{-1}, or $1/4\pi\epsilon_0 = 9 \times 10^9$ m F^{-1} approximately)

1 A capacitor has a charge of 20 μC (microcoulomb) when a p.d. of 200 V is applied to it. Calculate the capacitance of the capacitor. What is the charge on the capacitor if a battery of 40 V is connected?

2 Define *capacitance, microfarad*. Calculate the p.d. across a 2 μF capacitor if it has a charge of 80 μC. Calculate the new p.d. if the capacitor is then connected to an uncharged capacitor of 4 μF. What is the charge on each capacitor in this case?

3 A capacitor C is charged by a 20 V supply, and then discharged, at the rate of 100 times per second. If the current obtained is 80 μA, find C.

4 A capacitor of 4×10^{-4} μF is charged 100 times per second by a 20 V d.c. supply by means of a vibrating reed switch and discharged at the same rate through a galvanometer. What current in microampere is obtained?

5 A sphere of radius 6×10^{-2} m has a charge of 4×10^{-9} C. What is the potential of the sphere? What is its energy?

6 Define *dielectric constant* (or *relative permittivity*) of a medium. A parallel-plate capacitor has two square plates of side 0.1 m and are 20 mm apart. Find the capacitance when paraffin-waxed paper of dielectric constant 2 fills the space between the plates. Calculate the charge on a multiplate capacitor of 57 plates with the same dimensions, spacing, and dielectric, if a battery of 600 V is connected to the capacitor.

7 Two concentric spheres of radii 5.4 and 5.0 cm respectively form a capacitor, with a liquid between them of dielectric constant (relative permittivity) 3. Calculate the charge on the capacitor, and its energy, when a 300 V battery is connected to it.

With the battery disconnected, what is the new energy when the liquid is allowed to drain away?

8 By means of a vibrating reed switch, an air capacitor is charged 50 times per second by a 9 V d.c. supply and discharged at the same rate

ELECTRICITY 89

through a galvanometer. The capacitor consists of two square plates, each with sides 25 cm long and separated by a distance of 1 mm. If the deflection in the galvanometer is 5 mm, and the galvanometer sensitivity is 20 mm per microampere, calculate a value for ϵ_0, the permittivity of air.

9 Calculate the combined capacitance of (i) 2 μF and 3 μF capacitor in series, (ii) a 4 μF capacitor in series with a parallel arrangement of a 3 and 2 μF capacitor. Prove from first principles the formula for the combined capacitance of two capacitors in series and in parallel.

10 An isolated metal sphere X of radius 2 cm has a charge of 4×10^{-9} C, and is then connected to an isolated uncharged sphere of radius 8 cm. Calculate the initial potential of X, and the final potential and charge on X.

11 A capacitor of 2 μF is charged by a 100 V battery. Calculate the energy in the capacitor. If the capacitor is disconnected from the battery and then connected to a 6 μF uncharged capacitor, find the new energy in each capacitor. Account for the loss in energy which has occurred.

12 A capacitor of 4 μF is (i) in parallel, (ii) in series with a 6 μF capacitor, and a battery of 120 V is connected across the arrangement in each case. Calculate the charge on each capacitor, and the total energy of the capacitors in both cases.

13 Two capacitors of 25 μF and 100 μF respectively are joined in series with a d.c. supply of 6.0 V. What is the charge on each capacitor and the p.d. across each?

The supply is now disconnected without affecting the charge on each capacitor. Their two positive plates, and their two negative plates, are then connected together. Calculate (i) the common p.d. of the capacitors, (ii) the loss in energy of the two capacitors. How is this loss of energy accounted for?

14 (i) A parallel-plate air capacitor has two plates each of area 10^{-1} m^2 and 4 mm apart. Calculate the energy stored when a p.d. of 20 V is applied to the plates.

(ii) With the plates still connected to the 20 V supply, their separation is reduced to 2 mm. What is the new value of the energy stored in the capacitor? How has the change in energy occurred?

15 (i) A parallel-plate air capacitor has a capacitance of 5×10^{-10} F when the distance between the plates is 4 mm, and a p.d. of 200 V is connected to the capacitor. Calculate the charge on each plate and the energy stored.

(ii) The connections to the supply are then removed without affecting the charge, and the distance between the plates is increased by 1 mm. Calculate the new value of the capacitance and of the energy stored.

Explain the change in energy which has occurred.

16 A 100 V supply is connected to a 4 μF capacitor in series with a 2 MΩ (2 million ohms) resistor. Find (i) the current I at the instant of switching on the supply, (ii) the final charge on the capacitor, (iii) the time taken to charge the capacitor assuming the mean value of the current during flow of charge is $I/2$.

17 A 25 μF capacitor, previously charged by a p.d. of 10 V, is discharged through a 2 MΩ (2 x 10^6 Ω) resistor. What is:
(i) the initial charge on the capacitor?
(ii) the rate at which the capacitor discharges at the commencement of the discharge?
(iii) the new rate of discharge at the instant when the capacitor has lost 60% of its original charge?

18 Obtain from first principles the energy of a capacitor of a capacitance C when a p.d. V is connected to it. If the p.d. supply is disconnected, and the capacitor is then joined to an uncharged capacitor of capacitance C, prove that the total electrical energy has now diminished. Why has the total energy diminished?

19 Describe an experiment to show how the capacitance of a parallel-plate capacitor depends on the common area of the plates, their distance apart, and the dielectric between the plates.

20 Describe an experiment to measure the capacitance of an unknown capacitor of about 1 μF if a known capacitor of 0.5 μF is available.

21 Calculate the capacitance of a parallel-plate capacitor with square plates of side 10^{-1} m and distance apart 4 x 10^{-3} m and a dielectric of relative permittivity 5 between them. If the charge on the capacitor is 10^{-8} C, what p.d. was applied?

22 An isolated sphere in air has a radius of 5 x 10^{-2} m and a potential of 200 V. Calculate the electric field intensity just outside the sphere. If the maximum intensity is 3 x 10^6 V m^{-1}, what is the maximum potential?

23 A capacitor has square plates of side 2 x 10^{-2} m with a dielectric of relative permittivity 4.0 and thickness 2 mm between them. If a charge of 10^{-9} C is obtained on the capacitor when a battery is connected, calculate (i) the battery e.m.f. (ii) the energy stored in the capacitor.

24 An isolated spherical conductor of radius 50 mm has a charge of +15 x 10^{-9} C, and another spherical conductor of radius 40 mm has a charge of +30 x 10^{-9} C. Calculate the total energy of the spheres. If the spheres are now connected by a thin wire, find (i) their final potential, (ii) the charge on each, (iii) the loss in total energy.

25 Two circular metal discs each have a radius of 20 cm and are separated by a paper of thickness 4 mm and dielectric constant 2. Calculate the charge on the plates, and the energy of the capacitor, when a battery of 100 V is connected across them.

26 A multiplate capacitor is formed with 11 square plates each with sides 10 cm. If the plates are separated by glass of thickness 2 mm and dielectric constant 3, find (i) the capacitance of the capacitor, (ii) the energy stored if a battery of 200 V is connected across it.

38. MULTIPLE CHOICE QUESTIONS – ELECTROSTATICS

For each of the questions 1–10, choose one statement from A to E which is the most appropriate.

1. In electrostatics,
   A the volt is the unit of energy,
   B the newton per coulomb is the unit of p.d.,
   C 1 newton per coulomb = 1 volt per metre,
   D the metre per volt is the unit of work,
   E electric intensity x distance = potential gradient.
2. A glow may sometimes be observed round a high voltage spherical metal terminal because
   A the sphere is perfectly smooth,
   B the surrounding air becomes charged to the same voltage as the sphere,
   C electrons leak away to the air from the metal,
   D some parts of the sphere may be rough or pointed,
   E the air becomes non-conducting.
3. A van de Graaff generator model has two sets, X and Y, of pointed conductors in order that
   A X can transfer charges by contact with the moving belt,
   B X can spray the belt with charges and Y can transfer them by induction,
   C Y can transfer the charges by contact with the belt,
   D X can transfer the charges to Y directly,
   E X and Y can both induce charges on the high voltage terminal.
4. A charge of 2 microcoulombs is situated in a field of intensity 400 N C^{-1}. The force on the charge is
   A 8×10^{-4} N, B $8\pi \times 10^{-3}$ N, C 8π N, D 2000 N, E 8000π N.
5. When a small test charge is moved from one point to another in an electric field, the work done is
   A never zero,
   B measured in joule per metre,
   C independent of the path taken,
   D zero along the direction of the field,
   E measured in volt per metre.
6. Two parallel plates have a p.d. of 1000 V and are separated by 5 mm. When a small charge of 1 μC moves normally to the plates from one plate to the other, then the work done is
   A 500 J, B 2 J, C 0.5 J, D 5×10^{-3} J, E 10^{-3} J.
7. A hollow charged spherical conductor having a charge Q and radius r,
   A has a potential outside the sphere equal to $Q/4\pi\epsilon_0 r$,
   B has a potential at the surface equal to Q,
   C has an intensity inside equal to $Q/4\pi\epsilon_0 r^2$,
   D has zero intensity inside,
   E has flux inside which all passes through the centre.
8. In electrostatics:

A the potential gradient is zero where the electric intensity is zero,
B the potential is zero on the surface of a charged conductor,
C the potential gradient is a measure of the energy in the field,
D electric potential is a vector quantity,
E the potential is always zero in a vacuum although charges are present.

9 A parallel plate capacitor is joined to a battery. If the plates are moved further apart.
A no charge flows,
B some charge is returned to the battery,
C the p.d. between the plates is reduced,
D the energy in the capacitor increases,
E the energy in the battery is reduced.

10 A battery of e.m.f. E is connected to a capacitor in series with a resistor R. When the circuit is made:
A the current rises to a constant value,
B the p.d. across the capacitor decreases,
C the p.d. across the resistor increases,
D the capacitor charges initially at a rate given by E/R,
E the initial current is zero.

11 In the following diagram L, M, N are point charges which are equal in magnitude. L and M are fixed but N is free to move. POQ is the perpendicular bisector of LM, and N is placed on this line.

State what happens to N when
(a) all charges are positive,
(b) L and M are positive but N is negative,
(c) L and N are positive but M is negative,
Choose your answers from
A N accelerates to the right,
B N accelerates to the left,
C N starts to move up perpendicular to PQ,
D N starts to move down perpendicular to PQ,
E N oscillates about O.

12 Capacitors X of 10 μF and Y of 2μF are each charged to a p.d. of 10 V, disconnected from the charging battery and connected together. What happens when,
(a) the +ve plate of X is connected to the +ve plate of Y and the $-$ve plate of X is connected to the $-$ve plate of Y,
(b) the +ve plate of X is connected to the $-$ve plate of Y and the $-$ve plate of X is connected to the +ve plate of Y.
Choose your answers from
A Nothing,
B p.d. across both becomes greater than 10 V,
C p.d. across both becomes less than 10 V but is not zero,

D p.d. across both becomes zero,
E X charges up Y.

WORKED EXAMPLES ON ELECTROSTATICS

1 An electron of mass 9×10^{-31} kg and a charge of -1.6×10^{-19} C moves between two points in empty space having a p.d. of 500 V. Calculate the velocity of the electron arriving at the second point if it has zero velocity at the first point.

Work done on electron = charge (e) × p.d. (V)

∴ kinetic energy gained = $\frac{1}{2}mv^2$, where v is the velocity, $= eV$

$$\therefore v = \sqrt{\frac{2eV}{m}}$$

$$= \sqrt{2 \times \frac{1.6 \times 10^{-19}}{9 \times 10^{-31}} \times 500}$$

$$= 1.3 \times 10^7 \text{ m s}^{-1}$$

2 The maximum electric intensity outside the surface of an isolated spherical conductor is 3×10^6 V m^{-1}. Calculate the maximum charge on the sphere if its radius is 0.2 m and the maximum energy stored.

(i) At surface of sphere of radius r, electric intensity $= \dfrac{Q}{4\pi\epsilon_0 r^2}$

So $$\frac{Q}{4\pi \times 8.85 \times 10^{-12} \times 0.2^2} = 3 \times 10^6$$

$$\therefore Q = 3 \times 10^6 \times 4\pi \times 8.85 \times 10^{-12} \times 0.2^2$$

$$= 1.3 \times 10^{-5} \text{ C}$$

(ii) Capacitance of isolated sphere, $C = 4\pi\epsilon_0 r$

So $$\text{energy} = \frac{Q^2}{2C}$$

$$= \frac{(1.3 \times 10^{-5})^2}{2 \times 4 \times 8.85 \times 10^{-12} \times 0.2}$$

$$= 3.8 \text{ J}$$

3 A $2\mu F$ capacitor is charged by a battery of 100 V, and the plates are then connected to similarly charged plates of a $1 \mu F$ capacitor charged by a battery of 50 V. Calculate the final p.d. and charge of the first capacitor.

The charge on the $2 \mu F$ capacitor $= CV = \dfrac{200}{10^6}$ C; the charge on the $1 \mu F$ capacitor $= \dfrac{50}{10^6}$ C. Hence the total charge $= \dfrac{250}{10^6}$ C.

Suppose the final p.d. is V. Then the final total charge $= \left(\dfrac{2+1}{10^6}\right)V$.

94 GRADED EXERCISES AND WORKED EXAMPLES IN PHYSICS

$$\therefore \frac{3V}{10^6} = \frac{250}{10^6}$$

$$\therefore V = 83\tfrac{1}{3} \text{ V}.$$

$$\therefore \text{ charge on first capacitor} = CV = \frac{2}{10^6} \times 83\tfrac{1}{3} = \frac{166\tfrac{2}{3}}{10^6} \text{ C}$$

$$= 167 \ \mu\text{C}$$

4 A parallel-plate air capacitor has circular plates of 10 cm diameter 2 mm apart, and a p.d. of 300 V is connected between the plates. Calculate the energy of the capacitor, the surface-density of charge, and the electric intensity between the plates. ($\epsilon_0 = 8.85 \times 10^{-12}$ F m^{-1}.)

(1) $$C = \frac{\epsilon_0 A}{d} = \frac{8.85 \times 10^{-12} \times \pi \times (5 \times 10^{-2})^2}{2 \times 10^{-3}}$$

$$= 3.5 \times 10^{-11} \text{ F (approx.)}$$

(2) $$W = \tfrac{1}{2} CV^2 = \tfrac{1}{2} \times 3.5 \times 10^{-11} \times 300^2 \text{ J}$$

$$= 16 \times 10^{-7} \text{ J}$$

(3) $$Q = CV = 3.5 \times 10^{-11} \times 300 \text{ C}$$

$$\therefore \sigma = \frac{Q}{A} = \frac{3.5 \times 10^{-11} \times 300}{\pi \times (5 \times 10^{-2})^2}$$

$$= 1.3 \times 10^{-7} \text{ C m}^{-2}$$

(4) $$\text{Intensity} = \frac{V}{d} = \frac{300 \text{ V}}{2 \times 10^{-3} \text{ m}} = 1.5 \times 10^5 \text{ V m}^{-1}$$

5 An isolated sphere of radius 4 cm has a charge of 3×10^{-9} C, and is connected to another isolated uncharged sphere of radius 2 cm by a fine wire. Calculate the final potential and charge on the larger sphere, and its loss in energy. (Assume $1/4\pi\epsilon_0 = 9 \times 10^9$ F m^{-1}.)

The capacitance of an isolated sphere = $4\pi\epsilon_0 r$, where r is its radius in metre.

The total charge remains constant when the two spheres are connected. Suppose their final common potential is V. Then, since $Q = CV$, total charge after connection = $4\pi\epsilon_0 (0.04 + 0.02) V$.

$$\therefore \frac{1}{9 \times 10^9} \times 0.06 \ V = 3 \times 10^{-9}$$

$$\therefore V = 450 \text{ V}$$

The charge, Q, on the larger sphere = $CV = 4\pi\epsilon_0 \times 0.04 \times 450 = 2 \times 10^{-9}$ C.

The initial energy of the larger sphere = $\tfrac{1}{2}\dfrac{Q^2}{C} = \tfrac{1}{2} \times \dfrac{(3 \times 10^{-9})^2}{4\pi\epsilon_0 \times 0.04}$

$$= 1.0 \times 10^{-6} \text{ J}$$

The final energy $= \frac{1}{2}CV^2 = \frac{1}{2} \times 4\pi\epsilon_0 \times 0.04 \times 450^2 = 4.5 \times 10^{-7}$ J

∴ loss of energy $= 1.0 \times 10^{-6} - 4.5 \times 10^{-7} = 5.5 \times 10^{-7}$ J

CURRENT ELECTRICITY

39. CIRCUIT CALCULATIONS. CONDUCTORS

1 Calculate (i) the p.d. across a 25 Ω resistor carrying a current of 6 mA, (ii) the resistance of a wire carrying a current of 2 A when the p.d. across it is 0.4 V, (iii) the current in a 10 Ω resistor when the p.d. applied is 2 V.

2 A resistor has a p.d. of 40 V when the current in it is 120 mA. Find the p.d. across the resistor when the current is 0.6 A.

3 A resistor of 85 Ω is in series with a parallel arrangement of 60 and 20 Ω, and a p.d. of 120 V is connected across the whole circuit. Calculate (i) the current in each resistor, (ii) the p.d. across each.

4 A current of 9A flows towards the junction of a parallel arrangement of 6, 8 and 12 Ω. Find the current in each wire.

5 A battery of e.m.f. 12 V and internal resistance 4 Ω is connected across a 16 Ω resistor. Find (i) the current, (ii) the terminal p.d.

6 A battery of e.m.f. 6 V is connected across a 10 Ω resistor. If the p.d. across the resistor is 5 V, find (i) the current in the circuit, (ii) the internal resistance of the battery.

7 A battery of e.m.f. 12 V and internal resistance 8 Ω is connected to (i) a 4 Ω wire, (ii) a parallel arrangement of 6 and 3 Ω. Calculate the current through the battery, and the terminal p.d., in each case.

8 A battery of e.m.f. 4 V and internal resistance 2 Ω is connected in series with a battery of e.m.f. 6 V and internal resistance 3 Ω so that the e.m.f.s of the cells assist each other. A resistor of 15 Ω is joined to the two outer terminals. Find (i) the current flowing, (ii) the p.d. across the terminals of each cell.

9 In Qn. 8, one of the cells is reversed. Calculate the new current flowing.

10 Two batteries, each of e.m.f. 10 V and internal resistance 4 Ω, are joined in parallel with like poles connected together. An 8 Ω resistor is joined to the terminals. Find (i) the current in the resistor, (ii) the current through each cell, (iii) the p.d. across the terminals of the cells.

11 A 10 Ω and 6 Ω resistor are joined in series with batteries of respective e.m.f.s 4 V and 2 V, and internal resistances 1 Ω and 3 Ω. Fig. 39A. The junction of the two resistors is earthed. Find (i) the current flowing, (ii) the terminal p.d. of each battery, (iii), the potentials of the ends of the 10 Ω and 6 Ω resistors not earthed.

Fig. 39A.

12 State *Ohm's Law*. Why cannot a moving-coil voltmeter be used to verify the law? Describe an experiment to verify Ohm's law.

13 The $I-V$ characteristics in Fig. 39B are due respectively to (i) Nichrome wire, (ii) a diode valve, (iii) a gas, (iv) a junction (semi conductor) diode. Identify each characteristic.

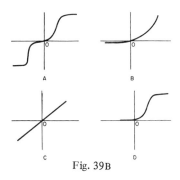

Fig. 39B

14 Name the charge carriers when current flows in (i) Nichrome wire, (ii) germanium, (iii) dilute copper sulphate solution, (iv) tungsten wire, (v) diode valve, (vi) junction diode. Which of these conductors would you class as *ohmic* or *non-ohmic* conductors, and why?

15 Explain the difference between the *drift velocity* and *thermal velocity* of an electron when current flows in a metal.

16 A current of 2 A flows through a copper wire of cross-section area 1 mm². If the number of free electrons per unit volume of copper is 8×10^{28} m^{-3}, calculate the drift velocity.

17 In Fig. 39C (i), the resistance R is varied until no current flows in the galvanometer G. Calculate R. What current flows through R?

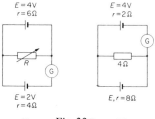

(i) Fig. 39C. (ii)

18 In Fig. 39C (ii), no current flows in the galvanometer G. Calculate the e.m.f. E of the lower battery and the current flowing from it.

40. ELECTRICAL ENERGY AND POWER

1 An electric lamp is rated at 60W–240V. Calculate the filament resistance when the lamp is operating normally. Find the cost of using 20 of the lamps continuously for a week if the Board of Trade unit costs 2p. What energy in megajoule (MJ) is used in this time?

2 A battery of 20 V and internal resistance 10 Ω is connected to a resistor of 40 Ω. Calculate the power developed in the resistor and in the battery, and find the efficiency of the arrangement.

3 Four 120 W–240 V and six 60 W–240 V lamps are used on a 240 V mains. Calculate the current flowing through the mains, the respective resistances of the 120 W and 60 W lamps when working, and the total energy used in megajoule (MJ).

4 A coil of 10 Ω is placed inside a calorimeter of mass 120 g and specific heat capacity 400 J kg^{-1} K^{-1} containing 80 g of water at 15°C. A battery of 10 V is connected to the coil. Calculate the final temperature of the water if the current flows for 2 min. (Specific heat capacity of water = 4200 J kg^{-1} K^{-1}.)

5 A heating coil is immersed in 200 g of aniline contained in a calorimeter of mass 100 g and specific heat capacity 400 J kg^{-1} K^{-1}. When the p.d. across the coil is maintained at 8 V and the current is 2 A, the aniline temperature rises from 10°C to 20°C in 5 min. Calculate the specific heat capacity of aniline.

6 Derive *from first principles* the formula for the heating effect of a current in terms of current I, resistance R, and time t. Draw a sketch of a hot-wire ammeter, and explain (i) its action, (ii) why it has a non-uniform scale.

7 A generator produces 100 kW at 500 V. If the power loss in copper cables 800 m long, connected to the generator, is 6% of this power, calculate the diameter of the cable. (Resistivity of copper = 1.6×10^{-8} Ω m.)

8 A 50 h.p. engine drives a dynamo of efficiency 80 per cent when working at full load. Calculate the current supplied by the dynamo at 500 V. (1 h.p. = 746 W.)

9 An iron clock-weight is 14 cm in diameter and 30 cm long. If the motor lifting the weight has an efficiency of 60 per cent, and the weight is raised 18 m in 10 s, find the power of the motor in kilowatts. (Density of iron = 7500 kg m^{-3}; assume $\pi = 22/7$.)

10 A battery of e.m.f. 12 V and internal resistance 8 Ω is connected to a resistance of R Ω. Calculate the power developed in the resistance. By using values of R from zero to 16 Ω, and then drawing a power against

R, determine the magnitude of R when the battery delivers *maximum* power to it.

41. WHEATSTONE BRIDGE. RESISTANCE.

1 A metre bridge has an unknown resistance X in one gap, and a resistance Y of 10 Ω in the other gap. The length of wire to the balance-point from the end of X is 37.5 cm. Calculate the resistance of X.

2 List the important practical points to be observed in measuring resistance accurately by a metre bridge.

3 In Question 1, Y is replaced by an unknown resistance Z and X by a known resistance of 5 Ω. Calculate Z if the balance-point is 42.8 cm from the end of the known resistance.

4 Two equal resistances are placed in series in one gap of a metre bridge, and a resistance of 20 Ω is placed in the other gap. The balance-point is then 40.0 cm from the end corresponding to the 20 Ω. Find the new position of the balance point from the 20 Ω end if the two resistances are now placed in parallel.

5 Derive from first principles the relation between the four resistances when a Wheatstone bridge circuit is balanced.

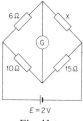

Fig. 41A

6 In the circuit of Fig. 41A, the battery of e.m.f. 2 V has negligible internal resistance. If no current flows in G, (i) calculate X, (ii) find the current in X, (iii) find the p.d. across the 15 Ω resistor.

Resistivity

7 Calculate the resistivity of a wire of resistance 5 Ω if its length is 2 metres and its diameter is 0.2 mm.

8 What length of eureka wire of diameter 0.1 mm is required to make a coil of 20 Ω? (Resistivity of eureka = 49 x 10^{-8} Ω m.)

9 Calculate the resistance of 10 m of manganin wire of diameter 0.36 mm. (Resistivity of manganin = 44 x 10^{-8} Ω m.)

10 The p.d. across 5 m of Nichrome wire is 3.2 V when the current in it is 0.2 A. Find the diameter of the wire, if the resistivity of nichrome is 10^{-6} Ω m.

Temperature Coefficient

11 Define *temperature coefficient* of a material. Calculate the temperature coefficient of platinum if a platinum wire has a resistance of 1.36 Ω at 0°C and a resistance of 1.89 Ω at 100°C. At what temperature will the resistance be 2.30 Ω?

12 A copper wire has a resistance of 2.46 Ω at 15°C and a resistance of 2.88 Ω at 70°C. Find the temperature coefficient of copper and the resistance of the wire at 100°C.

13 Name a material with a negative and a positive temperature coefficient. Describe fully, with circuit diagram, an experiment to measure the temperature coefficient of iron.

14 What do you know about the temperature coefficients of (i) manganin or eureka, (ii) copper, (iii) the semiconductors germanium or silicon? Draw a sketch of the winding of a resistance coil inside a box of standard resistances and name a material from which it is made.

15 A 0–15 mA meter of 10 Ω resistance is shunted to read 0–1.5 A. (i) Calculate the shunt resistance. (ii) Assuming the shunt has its correct value at 0°C, and its temperature is raised 10°C by flow of current, calculate the percentage error in the meter reading if the meter resistance itself is unaffected. (Temperature coefficient of shunt material = $4 \times 10^{-3} \mathrm{K}^{-1}$.)

42. POTENTIOMETER

1 In a potentiometer experiment, the length of wire giving a balance with a standard cell of 1.08 V is 53.4 cm, and with a cell X the length is 72.6 cm. Calculate the e.m.f. of X. Draw a complete circuit sketch of the arrangement.

2 A standard cell of 1.0186 V gives a balance-length at 51.2 cm on a potentiometer. If the p.d. across a standard 10 Ω coil in a circuit gives a balance-length at 78.4 cm, calculate the current in the circuit. Draw a circuit sketch of the two circuit arrangements required in the above case.

3 Give the theory of the potentiometer method of measuring the e.m.f. of a cell. What is the advantage of the potentiometer method over a voltmeter method of measuring e.m.f.?

4 In a potentiometer experiment, give a reason for using (i) an *accumulator* connected to the wire, (ii) a *uniform* wire, (iii) a *series resistance* with the galvanometer.

5 In the circuit of Fig. 42A, the cell of e.m.f. 2 V has negligible internal resistance and the potentiometer wire AB has a length of 100 cm and resistance 4 Ω.
(i) When $R = 1$ Ω, at what distance from A is the cell X 'balanced' if it has an e.m.f. E_1 of 1.0 V?
(ii) If X is replaced by another cell of e.m.f. 0.8 V, what value of R will produce a balance 50 cm from A?

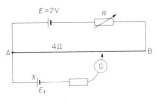

Fig. 42A.

6 Describe how two low resistances can be compared by means of the potentiometer, giving a complete circuit sketch. Why is a metre bridge unsuitable for this measurement?

7 The p.d. across the terminals of a cell on open circuit give a balance-length of 74.2 cm on a potentiometer. When a resistance of 15 Ω is connected to the cell, the terminal p.d. then gives a balance-length of 46.8 cm. Calculate the internal resistance of the cell. Prove any formula used.

8 The resistance of a potentiometer wire is 5 Ω and the wire is 1 metre long. What resistance is required, with a 2 V accumulator, in order that a p.d. of 1 microvolt per mm may be obtained? Describe how the arrangement is used to measure the e.m.f. of a thermocouple, giving a circuit diagram.

9 Describe and explain an experiment to measure the internal resistance of a cell by a potentiometer method.
State *two* precautions you should observe when doing the experiment.

10 Describe briefly how a potentiometer is used to compare the resistivities of two metals, given a length of each of about 4 Ω resistance.

11 Draw a circuit sketch showing respectively how (i) an ammeter (ii) a voltmeter can be calibrated by a potentiometer. Explain the method in each case.

12 In a potentiometer experiment to measure the internal resistance r of a cell, the terminal p.d. V is measured for different values of external resistance R, when current flows. Explain why a graph of $1/V$ against $1/R$ enables r to be found. Illustrate your answer from a rough sketch of the graph.

43. ELECTROLYSIS. CELLS

Electrolysis

1 A current of 1.1 A is passed through a copper sulphate solution for half an hour, and the mass of copper deposited is 0.66 g. Calculate the mass of copper deposited per coulomb.
Using Faraday's second law, what mass of silver is deposited per coulomb in the same circuit if the relative atomic masses of copper and silver are 63.5 and 108 respectively?

2 An ammeter is placed in series with a silver voltameter and reads 1.05 A. If 0.792 g of silver is deposited in 12 min, calculate the error in the ammeter. (Mass of silver deposited per coulomb = 1.1×10^{-6} kg C^{-1}.)

3 State *Faraday's two laws of electrolysis*. A cell, with a positive copper pole and a negative zinc pole, supplies a current of 0.1 A for 15 min. If 0.03 0 g of copper is then deposited, calculate the mass of copper deposited per coulomb and the mass of zinc lost. (Relative atomic masses of copper and zinc = 63.5 and 65.5 respectively).

4 A copper voltameter with copper electrodes has a resistance of 5 Ω, and is in parallel with a water voltameter with platinum electrodes having a resistance of 10 Ω. A 2 V accumulator is placed across the arrangement. Calculate the current from the accumulator if the back e.m.f. of polarization between hydrogen and oxygen is 1.1 V.

5 A water voltameter is placed in series with a silver voltameter and a copper voltameter. If 250 cm^3 of oxygen at 15°C and 750 mmHg pressure is collected, calculate the masses of hydrogen, silver and copper obtained. (Density of oxygen at s.t.p. = 0.72 kg m^{-3}; relative atomic masses of hydrogen, oxygen, copper, silver = 1, 16, 63, 108 respectively.

6 If the mass of silver deposited per coulomb is 1.12×10^{-6} kg C^{-1}, and the mass of copper deposited per coulomb is 3.28×10^{-7} kg C^{-1}, calculate the quantity of electricity required to deposit one mole of silver and of copper. Account for your answer, and explain its significance. (Relative atomic masses of silver and copper = 108 and 63.5 respectively and valencies = 1 and 2 respectively.)

7 Using the Faraday constant, 96 500 C mol^{-1} approximately, calculate the constant current needed to deposit 3.0 g of silver in 15 min. (Relative atomic mass of silver = 108, valency = 1.)

8 Use the following data to calculate the mass of a hydrogen atom:
Faraday constant = 96 500 C mol^{-1}.
Avogadro constant = 6.02×10^{23} mol^{-1}.
Mass of hydrogen deposited per coulomb = 1.05×10^{-8} kg C^{-1}.

9 Estimate the minimum back-e.m.f. in a water voltameter if 1 kg of hydrogen liberates 1.5×10^8 J on burning to form water and the mass of hydrogen deposited per coulomb is 1.04×10^{-8} kg C^{-1}.

102 GRADED EXERCISES AND WORKED EXAMPLES IN PHYSICS

10 Draw a circuit sketch showing how six accumulators are charged from the mains. How would you tell when the accumulators were fully charged? List the precautions adopted in the care and maintenance of accumulators.

11 Ten discharged accumulators have each an e.m.f. of 1.8 V and an internal resistance of 0.02 Ω. The accumulators are charged by a 100 V supply, and the charging current required is 2.5 A. Calculate the initial resistance required in the circuit, the power expended by the mains, and the power wasted in the accumulators.

12 Describe fully an experiment to measure the *back e.m.f.* of a water voltameter. Draw a diagram showing the $I-V$ characteristic of the water voltameter.

44. FORCE ON CONDUCTOR IN MAGNETIC FIELDS

1 (i) The *weber*, symbol Wb, is a unit of . . ., (ii) the *tesla*, symbol T, is a unit of . . ., (iii) 1 T = 1 Wb . . . . Fill in the missing words in (i), (ii), (iii).

2 A vertical wire carrying a current of 5 A is situated in a uniform horizontal magnetic field of 4×10^{-3} T. Calculate the magnitude of the force on the wire if the length of the wire is 0.4 m. Draw a sketch showing the direction of the force on the wire when the current flows (i) upwards, (ii) downwards.

3 Calculate the force on a wire of length 0.1 metre carrying a current of 10 A if it is (i) perpendicular, (ii) inclined at 30°, (iii) parallel to a field of 10^{-2} T. Show in a sketch the direction of the force in each case.

4 A rectangular coil of 20 turns and dimensions 4 cm by 2 cm is suspended with its plane and longer side vertical in a horizontal field of 2×10^{-2} T. If a current of 2 A flows in the coil, calculate the torque or moment of the couple initially on the coil when its plane is (i) parallel to the field, (ii) inclined at 60° to the field.

5 State *Fleming's rule* for the force on a current-carrying conductor in a magnetic field. Two long parallel wires X and Y each carry a current of 10 A and are 5 cm apart. Calculate (i) the flux density at Y due to the current in X, (ii) the force per metre on Y.

In a diagram, show the direction of the force when the currents are in (i) the same direction, (ii) the opposite direction.

6 Draw a labelled diagram of a *moving-coil ammeter*. What practical features produce (i) the uniform scale, (ii) high sensitivity? When a current I flows in the meter, a deflection θ is obtained. What is then (i) the deflecting couple on the coil, (ii) the opposing couple? Why is $I \propto \theta$?

7 Explain, with necessary calculations, how a 0–5 mA instrument of 10 Ω resistance is converted to an instrument reading (i) 0–1.5 A, (ii) 0–10 V, (iii) 0–3 A, (iv) 0–2 V.

8 What factors affect the *sensitivity* of a moving-coil meter? Draw a labelled diagram of a moving-coil mirror galvanometer. Describe how you would measure its sensitivity.

9 In one form of galvanometer, the sensitivity is 20 mm μA^{-1} and the mirror is effectively 1 metre from the scale. If the coil has an area of 1 cm^2 and 100 turns, and is situated in a radial field of 0.5 T, calculate the torsional constant of the instrument in newton-metre per radian.

10 A long solenoid with 3000 turns per metre carries a current of 4A. A horizontal straight wire X 4 cm long is in the middle of the coil perpendicular to its axis and also carries 4 A. Find the force on X and draw a diagram showing its direction. Suggest a design for a current balance based on this principle.

11 Explain how a 0–120 μA (microampere) instrument of 500 Ω resistance can be converted to (i) a 0–3 V instrument, (ii) a 0–1.2 mA instrument.

12 Draw a labelled diagram of a *moving-coil loudspeaker*, and explain its action. What factors may produce distortion?

13 A moving-coil instrument has a rectangular coil of 10 turns and dimensions 5 cm by 2 cm situated in a radial field of 0.4 T. The coil is suspended by a torsion wire which has a restoring couple of 2×10^{-6} newton-metre per degree of twist. Calculate the deflection of the coil when a current of (i) 1.2 mA, (ii) 40 μA is passed into it. What is the sensitivity of the instrument?

14 Draw a diagram of the magnetic flux pattern between two parallel very long straight current-carrying wires when the currents are (i) in the same direction, (ii) in opposite directions.

15 Draw a diagram of an *ampere-balance*. Show on it the direction of the current in the coils and the force on the moveable coil. Write down an expression for the current in terms of the force.

16 A ballistic galvanometer has no damping, whereas a current galvanometer has damping. Explain the reason for this, and state how these conditions are provided in the respective instruments. What other special feature of the moving system is needed in a ballistic-type galvanometer?

17 A moving-coil meter X has a coil of 50 turns of resistance 10 Ω. A similar meter Y has a coil of 60 turns of resistance 20 Ω. Which meter has (i) greater current sensitivity, (ii) greater voltage sensitivity, assuming X and Y have similar coil dimensions, magnetic fields and springs. Give reasons for your answers.

18 Draw diagrams of (i) a shunt-wound motor, (ii) a series-wound motor. Describe how the speed of each type of motor varies with the load, and explain your answer. Which motor would you use (i) for driving a tool machine, (ii) for raising a heavy load by a crane?

19 A horizontal strip of aluminium carries a current and is situated in a vertical magnetic field B. By considering the force on the moving electrons due to B, show that a p.d. (Hall voltage) is produced transversely across the strip. Draw a diagram in illustration.

20 In Qn. 19, write down: (i) the force due to B on an electron of charge e moving with a drift velocity v, (ii) the intensity due to the Hall voltage V_H if the width of the strip is b, (iii) hence the force on the electron due to the Hall voltage. By equating (i) and (iii), find V_H in terms of B, v and b.

21 The current in a strip of copper is given by $I = nevA$, where A is the cross-sectional area of the strip and n is the number of free electrons per unit volume. (i) If d is the thickness of the strip and b is its breadth, express ev in terms of I, n, b, d. (ii) Show that the Hall voltage V_H when a field B is applied is given by BI/ned. (iii) Calculate V_H if $B = 1$ T, $I = 6$ A, $n = 7.5 \times 10^{28}$ m^{-3}, $e = 1.6 \times 10^{-19}$ C, and $d = 1$ mm.

45. ELECTROMAGNETIC INDUCTION

1 State *Faraday's law of electromagnetic induction*. A coil of 50 turns and area of cross-section 4×10^{-3} m^2 is situated in a field of 10×10^{-3} T so that the flux enters all the turns normally. If the flux density (i) diminishes to 2×10^{-3} T in 1/100 s, (ii) increases to 60×10^{-3} T in 1/20 s, calculate the average induced e.m.f. in each case.

2 A coil of 100 turns and cross-sectional area 2×10^{-3} m^2 is placed in a field of 8×10^{-3} T so that the flux enters all the turns normally. Calculate the average induced e.m.f. if the field is reversed in 1/50 s.

3 The S pole of a magnet is moved (i) towards, (ii) away from, (iii) right through a coil connected to a sensiive galvanometer. Draw sketches showing clearly the direction of the induced current in each case.

4 A primary coil, with an accumulator and switch in series, is placed near to and on the same axis as a secondary coil connected to a galvanometer. Draw sketches showing clearly the directions of the induced current when the switch is turned on and off. Explain your answers.

5 State *Lenz's law*. Describe how you would verify it experimentally. How does Lenz's law follow from the principle of the conservation of energy?

6 State *Fleming's rule* for the direction of the induced current in a straight wire moving through a magnetic field. A conductor of length 2 m moves (i) perpendicular, (ii) parallel to its length with a velocity of 1 m s^{-1}. If the conductor moves perpendicularly to a magnetic field of 2×10^{-5} T in each case, calculate the induced e.m.f. in the wire. Draw a diagram showing the direction of the e.m.f., and mark the 'positive' end of the conductor considered as a generator of e.m.f.

ELECTRICITY

7 A rail on a train moving due west is 1.5 m long and points northwards. If the speed of the train is 12 m s^{-1}, calculate the induced voltage in the rail. (Earth's horizontal component = 2 x 10^{-5} T, angle of dip = 70°.)

8 A vertical metal disc of radius 0.2 m is rotated at 5 rev s^{-1} about its centre in a horizontal uniform magnetic field of 0.1 T which is normal to the plane of the disc. Find the e.m.f. produced between the centre and the top of the disc. Explain any formula used and show the direction of the e.m.f. in a sketch.
What is the e.m.f. between the ends of a diameter of the disc?

9 A rectangular coil of 100 turns and area 4 x 10^{-2} m^2 is rotated about a horizontal axis at a constant rate of 50 rev s^{-1} in a horizontal magnetic field of 0.2 T perpendicular to the axis. Calculate (i) the maximum value of the induced e.m.f. in the coil, (ii) the induced e.m.f. when the plane of the coil is 30° to the horizontal, (iii) the induced e.m.f. when the plane of the coil is vertical. Derive any formula you use from first principles.

10 Describe a simple a.c. generator, and explain with diagrams the changes in direction and magnitude of the e.m.f. during a complete cycle. Explain how direct current (d.c.) is obtained from this generator.

11 A coil of 20 turns and area 3 x 10^{-2} m^2 placed in a horizontal field of 10^{-2} T so that the flux enters all the turns normally. The coil has a resistance of 10 Ω and is connected to a ballistic galvanometer of 40 Ω resistance. Calculate the quantity of charge in microcoulomb through the galvanometer when the plane of the coil is (i) rotated to the horizontal, (ii) completely reversed. Derive any formula you use.

12 Explain why the coil of a moving-coil ammeter is wound on a metal frame, and the coil of a ballistic galvanometer is wound on a non-conducting frame. Why is the inertia of the moving system made large for the ballistic galvanometer?

13 What are *eddy currents*? (i) Describe an industrial application of eddy currents. (ii) Explain, with diagrams, how eddy current losses of energy are reduced in an iron-cored transformer.

14 Describe a *step-up* and *step-down* iron-cored transformer. Explain how they function. Show that, on open circuit, the ratio of secondary e.m.f. to primary e.m.f. = the ratio of secondary to primary turns. What effect is observed in the primary when the secondary circuit is closed, and why?

15 A coil of 50 turns and area 10^{-2} m^2 is placed with its plane normal to the field between the poles of a powerful magnet. The coil has a resistance of 5 Ω and is connected to a ballistic galvanometer of 35 Ω. When the coil is removed completely from the field, a deflection of 120 divisions is registered in the galvanometer. Calculate the flux density B in the field if the sensitivity of the galvanometer is 15 divisions per microcoulomb.

16 The flux linking a solenoid is 3 Wb when a steady current of 2 A flows. What is the inductance L of the coil? The area A of the iron core is 6×10^{-4} m², the length l of the coil is 0.2 m and the relative permeability μ_r of the iron is 100. Find the approximate number of turns N needed to make this inductance, assuming $L = \mu_0 \mu_r N^2 A/l$ approximately. ($\mu_0 = 4\pi \times 10^{-7}$ H m⁻¹).

17 A steady current of 2 A flows in a long solenoid with an air core. A small narrow coil of 50 turns and area 10^{-3} m² is placed in the middle of the solenoid and coaxially with it, and connected to a ballistic galvanometer. When the current is reversed in the solenoid, a charge of 10 microcoulombs circulates in the galvanometer. If the total resistance of galvanometer and coil is 5 Ω, calculate (i) the flux density in the middle of the solenoid, (ii) the approximate inductance of the solenoid if its area is 12×10^{-4} m² and it has 200 turns.

18 A coil X is placed near a coil Y. When a steady current of 4 A flows in X, the flux linking Y is 8×10^{-2} Wb. Calculate the mutual inductance between X and Y. What amount of flux will link X when a steady current of 2 A flows in Y?

19 A 20 V battery of negligible resistance is connected to a coil of inductance 5 H and negligible resistance in series with a 10 Ω resistor.

What is the rate of growth of the current (i) when the circuit is first made, (ii) when the current just reaches the value of 0.5 A?

Draw a sketch showing how the current varies with the time, and indicate on it the maximum value of the current.

46. MAGNETIC FIELDS OF CONDUCTORS

($\mu_0 = 4\pi \times 10^{-7}$ H m⁻¹)

1 Fill in the missing current-carrying conductors in each case:
(i) $B = \mu_0 I/2r$ for a . . . (ii) $B = \mu_0 I/2\pi r$ for a . . . (iii) $B = \mu_0 NI/l$ (approx.) for a . . ..

2 State *Maxwell's corkscrew rule*. Explain how it is applied to find the magnetic field direction in the cases of a straight and a circular current-carrying conductor. Draw diagrams of the field directions at points due east, west, north and south of a straight vertical current-carrying wire.

How does the field round a straight wire differ from that inside a current-carrying solenoid?

3 A current of 4 A flows down a straight vertical wire passing through a horizontal board. Calculate the flux density at a point 0.1 m from the wire on the board.

Draw a sketch of the flux pattern on the board, indicating the position of the 'neutral point'. If this point is 4 cm from the wire, calculate B_{hor}, the flux density of the earth's horizontal component.

4 Describe an experiment to show that the field value B outside a long straight current-carrying wire is inversely proportional to the distance from the wire, listing the measurements required.

5 Helmholtz coils are two coils of radius r which are distance r apart. Name an experiment in which these coils are used. Draw a sketch showing how the field B varies between the coils and point out a useful feature of the field.

6 A solenoid has 100 turns and is 0.5 m long. Calculate the approximate value of B in the middle of the solenoid when the current is 2 A.

What is the approximate force on a wire 1 cm long and carrying a current of 3 A when it is in the middle of the solenoid and perpendicular to its axis?

7 Describe an experiment to show that B is directly proportional to the current in a solenoid, and list the measurements needed. How would you use a 'Slinky' coil to show that $B \propto n$ for a given current, where n is the number of turns per metre?

8 Two parallel straight conductors X and Y are 0.1 m apart. X is a long conductor and carries a downward current of 8 A. Y is a short conductor 0.04 m long and carries a current of 5 A in the opposite direction to that in X. Calculate (i) the value of B at Y due to X, (ii) the approximate force on Y.

Draw a sketch of the arrangement and show the direction of the force on Y.

9 The force per metre between two infinitely-long straight conductors, each carrying a current of 1 A and separated a distance of 1 m in a vacuum of permeability μ_0 is 2×10^{-7} N. Show that $\mu_0 = 4\pi \times 10^{-7}$ H m^{-1}.

47. MAGNETIC PROPERTIES OF MATERIALS

1 Draw typical *B-H* loops for soft iron and for steel on the same axis. Explain how the curves are used to show the differences between the magnetic properties of the two materials and state the differences.

2 Define the term *hysteresis*, and explain why it occurs. How is the hysteresis loss per cycle calculated from the *B-H* loop? How is the hysteresis loss in materials diminished in practice?

3 Define the terms *coercive force, remanence, susceptibility, permeability*. What magnetic properties are required in materials used for (i) transformer cores, (ii) permanent magnets, (iii) electromagnets, (iv) diaphragms of a telephone earpiece?

4 Using domain theory, explain with the aid of diagrams how ferromagnetic materials become magnetized. Why does 'magnetic saturation' occur?

5 *Paramagnetic, diamagnetic* and *ferromagnetic* materials are three different types of magnetic materials. With the aid of diagrams, describe their difference in behaviour in a magnetic field. What reasons can you give for these differences?

6 A soft-iron ring of average circumference 0.4 m has a primary coil of 200 turns wound all round it which carries a current of 1 A. A secondary coil wound round the ring at one place has five turns, and is connected to a ballistic galvanometer, the total resistance of this circuit being 20 Ω. When the current is reversed in the primary coil a deflection of 30 divisions is obtained in the galvanometer. Calculate the flux density in the iron at a current of 1 A and the relative permeability of the iron. (Area of cross-section of ring = 4×10^{-4} m^2; sensitivity of galvanometer = 0.5 division per microcoulomb.)

48. A.C. CIRCUITS

1 An alternating current is represented by $I = 5 \sin 200 \pi t$, where I is in A, t is the time in seconds. What is (i) the peak value of current, (ii) the r.m.s. value, (iii) the current flowing 1/1200 s after the current changes direction, (iv) the power dissipated in a non-inductive resistance of 4 Ω?

2 Repeat Question 1 for a current represented by $I = 8 \sin 100\pi t$.

3 Define *root-mean-square value* of current. How is it related to the peak value for a sine variation? Calculate the r.m.s. voltage across (i) a resistance of 100 Ω, (ii) a capacitor of 10 μF if a current if a current of 20 mA (r.m.s.), frequency 50 Hz flows in each. Draw a phasor diagram of the voltages.

4 Calculate the reactance of (i) an inductance of 2 H at 100 Hz, (ii) a 1 μF capacitor at 200 Hz.

5 An a.c. voltage of 100 V (r.m.s.), frequency 50 Hz, is connected to a resistance of 50 Ω and then to an inductance of 1 H. Calculate the current flowing in each case, and draw a phasor diagram of the voltage and current for each.

6 An alternating current of 0.1 mA (r.m.s.), frequency 250/2π Hz, flows through a resistance of 3000 Ω in series with a capacitor of 1 μF. Calculate the voltage across each component and across both components. Draw a phasor diagram of the voltage and current for each component.

7 A voltage of 10 V (r.m.s.), frequency 150/2π Hz, is connected across an inductance of 2 H and negligible resistance in series with a non-inductive resistance of 400 Ω. Calculate (i) the current flowing, (ii) the voltage across each component, (iii) the phase angle between the current and applied voltage, (iv) the power absorbed in the circuit.

8 Repeat Question 7 if the same voltage is connected across a capacitor of 4 μF in series with a resistance of 300 Ω, the frequency now being 250/2π Hz.

ELECTRICITY

9 What is meant by *reactance* of a coil? An a.c. voltage of frequency $400/2\pi$ Hz, is connected to a coil of resistance 100 Ω and inductance L in series with a capacitor of 2.5 μF. The voltage across the capacitor is 40 V (r.m.s.) and that across the coil is 160 V (r.m.s.). Calculate (i) the current flowing in the circuit, (ii) the impedance of the coil, (iii) the magnitude of L, (iv) the power dissipated in the coil.

10 Define *impedance* of a circuit. A coil of negligible resistance has a reactance of 5000 Ω to a frequency of $400/2\pi$ Hz and is connected to a resistance. Calculate the inductance of the coil. Calculate the capacitance of a capacitor which is placed in series in the circuit and does not alter the impedance of the circuit.

11 Calculate the current and the power dissipated in a coil of inductance 2 H and resistance 400 Ω when an a.c. voltage of 100 V (r.m.s.), frequency $400/2\pi$ Hz, is connected to it.

12 Describe an instrument for measuring alternating current, and explain how it functions.

13 Describe and explain how the root-mean-square value of a current in a circuit can be measured by a heating experiment, if no a.c. meter is available.

14 A varying voltage has a 'square' wave shape of peak value 2.0 V for half a cycle and zero voltage for the other half of the cycle. Calculate the r.m.s. value of the voltage.

15 A coil has an inductance of 0.02 H and resistance 50 Ω and is in series with a capacitor of 0.0002 μF. An alternating voltage of 0.2 V (r.m.s.) is connected at the resonant frequency f_0 of the circuit. Calculate (i) f_0, (ii) the current flowing, (iii) the voltage across the capacitor, (iv) the power absorbed in the circuit.

16 A coil, capacitor and a lamp are all in series with an a.c. supply. When the frequency of the supply is increased from a low value, the brightness of the lamp increases to a maximum and then decreases. (i) Explain why this happens, (ii) write down an expression for the frequency when maximum brightness is obtained, (iii) write down an expression for the magnitude of the maximum current obtained.

17 A coil of inductance 0.4 H and resistance 10 Ω is in series with a 0.1 μF capacitor C. When the frequency of an a.c. supply is varied and the magnitude V of the voltage is maintained constant, a maximum current of 0.5 A is obtained. Calculate (i) V, (ii) the voltage across C and across the coil at maximum current.

18 A capacitor of 10 μF and a coil of 2 H and negligible resistance are placed in *parallel* across a mains supply of 20 V and f = 50 Hz. Calculate the current in each component and draw a phasor diagram of the currents.

What is the current flowing from the mains supply?

WORKED EXAMPLES ON CURRENT ELECTRICITY

1 Two resistors of 2000 and 3000 Ω respectively are placed in series across a 200 V battery of negligible internal resistance. A moving-coil voltmeter is placed across each resistor in turn. What is the p.d. registered on the instrument in each case, if the voltmeter has a resistance of 6000 Ω?

When the 6000 Ω voltmeter is placed across the 2000 Ω resistor, their combined resistance R is given by

$$\frac{1}{R} = \frac{1}{2000} + \frac{1}{6000} = \frac{4}{6000}.$$

$$\therefore \quad R = \frac{6000}{4} = 1500 \ \Omega$$

$$\therefore \quad \text{current}, I, = \frac{200}{1500 + 3000} = \frac{2}{45} \ \text{A}$$

$$\therefore \quad \text{p.d. across voltmeter} = IR = \frac{2}{45} \times 1500 = 67 \ \text{V}.$$

This is the p.d. registered on the meter.

When the 6000 Ω voltmeter is transferred to the 3000 Ω resistor, their combined resistance S is given by

$$\frac{1}{S} = \frac{1}{3000} + \frac{1}{6000},$$

from which

$$S = 2000$$

$$\therefore \quad \text{current } I = \frac{200}{2000 + 2000} = \frac{1}{20} \ \text{A}$$

$$\therefore \quad \text{p.d. across voltmeter} = \frac{1}{20} \times S = \frac{1}{20} \times 2000 = 100 \ \text{V}.$$

This is the new p.d. registered on the meter.

(*Note.* The current in the circuit alters when the voltmeter is transferred from one resistor to the other.)

2 The coil of a mirror galvanometer has 20 turns and dimensions 5 cm by 2 cm, and is situated in a radial field of 0.4 T. If the restoring couple per degree is 10^{-5} N m, calculate the current in the instrument when the deflection is $0.04°$.

$$\text{Deflecting couple} = NABI = c\theta,$$

where $N = 20$, $A = 5 \times 2 \ \text{cm}^2 = 10 \times 10^{-4} \ \text{m}^2$, $B = 0.4$ T.

$$\therefore I = \frac{c\theta}{NAB}$$

$$= \frac{10^{-5} \times 0.04}{20 \times 10 \times 10^{-4} \times 0.4}$$

$$= 5 \times 10^{-5} \text{ A} = 50 \, \mu\text{A}$$

3 Name three substances which do not obey Ohm's law. A wire 1 m long and 1 mm^2 cross-section has a p.d. of 0.4 V across its ends when a current of 2 A flows in it, its temperature being then 15°C. Calculate the resistivity of the wire at 200°C if its temperature coefficient is 0.0040 K^{-1}.

First Part. A metal (copper oxide-copper) rectifier, a gas and dilute sulphuric acid between platinum electrodes, do not obey Ohm's law, i.e. the p.d., V, is not directly proportional to the current, I, at constant temperature.

Second Part. The resistance, R_1, of the wire at 15°C = $\dfrac{V}{I} = \dfrac{0.4}{2} = 0.2 \, \Omega$.

But
$$R = R_0(1 + \alpha t),$$

where R_0 is the resistance at 0°C, α is the temperature coefficient. Hence if R_2 is the resistance at 200°C,

$$R_2 = R_0(1 + 0.004 \times 200)$$

$$R_1 = 0.2 = R_0(1 + 0.004 \times 15)$$

$$\therefore \frac{R_2}{0.2} = \frac{1 + 0.004 \times 200}{1 + 0.004 \times 15}$$

$$\therefore R_2 = 0.2 \times \frac{1.8}{1.06}.$$

But
$$R_2 = \rho \frac{l}{A} = \rho \times \frac{1}{10^{-6}}$$

since $l = 1$ m, $A = 1$ mm$^2 = 10^{-6}$ m^2

$$\therefore \frac{0.2 \times 1.8}{1.06} = \rho \times \frac{1}{10^{-6}}$$

$$\therefore \rho = \frac{0.2 \times 1.8 \times 10^{-6}}{1.06 \times 1} \, \Omega \, \text{m}$$

$$= 3.4 \times 10^{-7} \, \Omega \, \text{m}$$

4 The rail of a train moving due west is 1 metre long, and the rail points due north. If the velocity of the train is 100 km per hour, calculate

the induced e.m.f. in the rail and state its direction. (Earth's horizontal component, $B_{hor} = 2 \times 10^{-5}$ T, angle of dip = 65°.)

The induced e.m.f., E, along the rail is due to the cutting of the flux due to the *vertical* component, B_v of the earth's field, since B_v is perpendicular to both the length of the rail and its direction of motion. The magnitude of E in volt is given by

$$E = B_v l v$$

where l is the length of the rail in metre, v is the velocity in metre per second, and B_v is in T.

Now

$$B_v = B_{hor} \tan 65° = 2 \times 10^{-5} \tan 65°,$$

$l = 1$ m, $v = 100\,000/3600 = 10^3/36$ m s^{-1}.

$$\therefore \quad E = 2 \times 10^{-5} \tan 65° \times 1 \times 10^{-3}/36$$
$$= 1.2 \times 10^{-3} \text{ V}.$$

5 A coil of inductance 2 H and resistance 50 Ω is in series with a resistor of 500 Ω, and an a.c. voltage of 80 V. $f = 50$ Hz, is connected across the arrangement. Calculate (i) the current flowing, (ii) the voltage across the coil, (iii) the power absorbed, (iv) the capacitance needed in series to produce the maximum current in the circuit, (v) the magnitude of the maximum current in (iv)

Reactance of coil inductance = $2\pi f L = 2\pi \times 50 \times 2 = 628$ Ω
Resistance of whole circuit = $50 + 450 = 500$ Ω

$\therefore$ impedance of whole circuit, $Z = \sqrt{X_L^2 + R^2} = \sqrt{628^2 + 500^2}$

(i) $\quad \therefore \quad I = \dfrac{V}{Z} = \dfrac{80}{\sqrt{628^2 + 50^2}} = 0.1$ A (approx.)

(ii) Impedance, Z, of coil = $\sqrt{628^2 + 50^2} = 630$ Ω

$\therefore \quad V = IZ = 0.1 \times 630 = 63$ V

(iii) Power is absorbed only in the circuit resistance.

$\therefore \quad P = I^2 R = 0.1^2 \times 500 = 5$ W

(iv) For resonance, $f_0 = 1/2\pi\sqrt{LC} = 50$ Hz

$\therefore \quad C = \dfrac{1}{4\pi^2 f_0^2 L} = \dfrac{1}{4\pi^2 \times 50^2 \times 2}$ F = 5×10^{-6} F = $5\,\mu$F

(v) Maximum current $I = \dfrac{V}{R} = \dfrac{80}{500} = 0.16$ A

6 A rectangular coil of area 20 cm^2 and 10 turns is placed between the poles of a magnet so that its plane is perpendicular to the flux, and the ends of the coil are connected to a ballistic galvanometer. The total resistance of the circuit is 10 Ω, and when the coil is removed a charge of 60 microcoulombs is measured by the galvanometer. Calculate the flux density in the magnet field.

$$Q = \frac{\Phi}{R} = \frac{NAB}{R},$$

where $Q = 60 \times 10^{-6}$ C, $N = 10$, $A = 20$ cm$^2 = 20 \times 10^{-4}$ m^2, $R = 10$ Ω

$$\therefore\ B = \frac{QR}{NA}$$

$$= \frac{60 \times 10}{10^6 \times 10 \times 20 \times 10^{-4}}$$

$$= 0.03\ \text{T}$$

7 A calorimeter of heat capacity 80 J K^{-1} contains water of mass 0.1 kg and a coil of 5 Ω resistance totally immersed in the water. The coil is placed in parallel with a copper voltameter having copper electrodes and a resistance of 7 Ω. When the arrangement is placed in a circuit, 0.66 g of copper is deposited in 40 min. Calculate the temperature rise of the calorimeter in the same time. (Specific heat capacity of water = 4200 J kg^{-1} K^{-1}, mass of copper deposited per coulomb = 3.3 × 10^{-7} kg C^{-1}.)

$$\text{Current in voltameter,}\ I = \frac{m}{t} = \frac{0.66 \times 10^{-3}}{3.3 \times 10^{-7} \times 40 \times 60} = \tfrac{5}{6}\ \text{A}$$

Since the coil (5 Ω) is in parallel with the voltameter (7 Ω),

$$\therefore\ \text{current in coil,}\ I, = \tfrac{7}{5} \times \tfrac{5}{6}\ \text{A} = \tfrac{7}{6}\ \text{A}$$

$$\therefore\ \text{Heat produced in joule} = I^2 Rt = (\tfrac{7}{6})^2 \times 5 \times 2400$$

Total heat capacity of water and calorimeter = 0.1 × 4200 + 80 = 500 J K^{-1}

$$\therefore\ \text{temperature rise} = \left(\frac{7}{6}\right)^2 \times \frac{5 \times 2400}{500}$$

$$= 33°\text{C (approx.)}$$

8
Atomic Physics

49. ELECTRONS AND IONS – MOTION IN ELECTRIC AND MAGNETIC FIELDS

(In all questions, assume where necessary that: electron charge $e = 1.6 \times 10^{-19}$ C, $e/m_e = 1.8 \times 10^{11}$ C kg^{-1}.)

1 The charge-mass ratio or *specific charge* (e/m_e) of an electron is 1.76×10^{11} C kg^{-1}; the mass-charge ratio (m_H/e) of a hydrogen ion is about 1.05×10^{-8} kg C^{-1}. Calculate the ratio of the mass of an electron to that of a hydrogen ion.

2 An electron is situated between two plates 20 mm apart having a p.d. of 600 V. Calculate the force on the electron and its initial acceleration in the field.

3 The force on a charge of $4e$ between two plates 32 mm apart is 8×10^{-14} N. Calculate the p.d. between the plates.

4 An electron beam with a velocity of 10^6 m s^{-1} enters a magnetic field of 4×10^{-5} T everywhere normal to the beam. Calculate the radius of the path of the beam.

5 A beam of ions, with a charge-mass ratio of 5.0×10^8 C kg^{-1}, enters a magnetic field B normal to the beam with a velocity of 10^5 m s^{-1}. The radius of the circular beam is 40 mm. Calculate B.

6 In an electron tube, electrons from the cathode start from rest and move to the anode through a p.d. of 150 V. Calculate the velocity of the electrons on reaching the anode.

If the electron beam passes through a hole in the anode, and enters a magnetic field of 10^{-4} T normal to the direction of the beam, calculate the radius of the circular path in the field.

7 An oil-drop of mass m carries a charge $20e$ and is held stationary between two plates having a p.d. of 300 V and 5 mm apart. Calculate the mass of the oil-drop and its radius, given the density of oil is 900 kg m^{-3} and $g = 10$ m s^{-2}.

8 A beam of protons, moving with a velocity of 5×10^6 m s^{-1}, is passed between two parallel plates 5 mm apart and 20 mm long in a direction initially parallel to the plates. A p.d. of 5 kV is applied between the plates. Find the deflection on a screen 300 mm away. (Assume e/m_H for protons $= 10^8$ C kg^{-1}.)

9 A current of 2 A flows along a copper wire of cross-sectional area

ATOMIC PHYSICS 115

1 mm^2. Assuming 10^{29} free electrons per m^3 in copper, find the drift velocity of the electrons.

10 An electron beam is accelerated by a p.d. of 300 V, and then enters normally a magnetic field of 10^{-3} T. Calculate the velocity attained by the electrons and the radius of its path in the magnetic field.

11 A charged oil-drop falls under gravity in air of viscosity 1.8 × 10^{-5} N s m^{-2} with a velocity of 4.5 × 10^{-4} m s^{-1} between two plates 5 mm apart. When a p.d. of 4500 V is connected between the plates, the drop rises with a velocity of 1.8 × 10^{-4} m s^{-1}. Calculate (i) the radius of the drop, (ii) the charge on the drop. (Density of oil = 900 kg m^{-3}; neglect the density of air and assume g = 10 m s^{-2}.)

12 In a Thomson experiment to determine the charge-mass ratio of an electron, the electron beam is undeflected when a magnetic field of 2 × 10^{-3} T is opposed by an electric field between two plates 10 mm apart and having a p.d. of 160 V. Find the velocity of the electrons. If the beam travels in a circular path of radius 23 mm in the uniform magnetic field alone, estimate the charge-mass ratio of an electron.

13 In a fine beam tube, electrons are first accelerated by an anode voltage V_a of 150 V and then deflected in a circle by a uniform magnetic field B of 10^{-3} T due to two Helmholtz coils. The radius of the circle is measured as 40 mm. Estimate a value for the charge-mass ratio (e/m_e) of an electron.

[*Hint*: $\frac{1}{2} m_e v^2 = eV_a$ (1), $Bev = m_e v^2 / r$ (2).]

14 The electron beam in a TV tube carries a current of 8 mA. How many electrons per second cross a particular section of the beam?

If the electrons are incident on the face of the tube after acceleration by a p.d. of 1000 V, calculate (i) the velocity of the electrons just before striking the tube, assuming they have zero initial velocity, (ii) the momentum change per second, or force, on impact with the tube, if the mass of an electron is 9 × 10^{-31} kg and the velocity after impact is zero.

15 Protons are whirled in a horizontal circle of radius r by an applied vertical field B. The charge-mass ratio of the proton is q/M and the velocity in the circle is v. Write down the relationship for the centripetal force on a proton, and hence find v in terms of q, M, r and B.

If $q/M = 10^8$ C kg^{-1} and B = 2 T, calculate the period of the circular motion and its frequency.

16 In Thomson's experiment to find the charge-mass (e/m_e) ratio of an electron, (i) draw a sketch of the apparatus, (ii) indicate on it the directions of the electric and magnetic fields, (iii) list all the measurements needed to find e/m_e. What assumption and additional information was needed to find the mass of the electron?

17 In Millikan's experiment to find the charge e on an electron, (i) draw a sketch of the apparatus, (ii) explain how a particular oil-drop obtained varying charges, (iii) list all the measurements required, (iv) write down the formulae required, (v) explain how the magnitude of e is obtained from all the charges measured.

18 An electron rotates in a circle of radius 5×10^{-12} m with a constant frequency of 10^{15} rev s^{-1}. Calculate (i) the current produced, (ii) the magnetic field value B at the centre of the circle.

50. DIODE VALVE. CATHODE RAY TUBE JUNCTION DIODE. TRANSISTOR

Diode valve

1 Describe, with circuit diagram, how the characteristic $(I - V)$ curve of a diode valve is obtained. How is the a.c. or slope resistance, r_a of the valve obtained?

2 Draw characteristic curves of a diode valve at two different cathode temperatures. Explain the shape of the curves.

3 Describe, with circuit details, how a diode valve converts a.c. to d.c. voltage. Why is a 'filter circuit' needed? Add a suitable filter circuit to your circuit if not already done.

4 'The current in a diode valve is first space-charge limited and then saturated.' Explain the meaning of this statement.

5 The following results were obtained with a diode valve:

V_a(V)	20	40	60	80	100
I_a(mA)	1.2	3.2	6.6	9.8	13.0

(i) Draw the characteristic and calculate the a.c. resistance.

(ii) If $I_a = kV_a^n$, where k and n are constants, plot $\log I_a$ v. $\log V_a$ and obtain n.

6 A circuit for converting a.c. mains to d.c. voltage consists of (i) double-diode valve, (ii) a transformer with a centre-tapping, (iii) a resistor as a load, (iv) a filter circuit consisting of a capacitor and an inductor.

(a) Draw a sketch of the circuit. (b) Explain fully why this is called a 'full wave' rectifier. (c) Explain briefly the purpose and action of the filter circuit.

Cathode Ray Tube

7 Draw a diagram showing the arrangement of the electrodes inside a typical cathode ray tube. Label the electrodes and explain their functions.

8 (i) The time base speed of an oscilloscope is 2.5 microseconds per mm. The separation between the transmitted and received pulses from a distant aircraft is 80 mm. Calculate the distance of the aircraft. ($c = 3 \times 10^5$ km s^{-1}.)

(ii) The time base of an oscilloscope is set on 0.5 milliseconds per mm. Six cycles of a waveform under study occupy 30 mm. Find the frequency of the waveform.

9 By means of Lissajous' figures, explain how you would use a cathode ray oscilloscope (i) to calibrate an oscillator using the mains frequency as a stable source of 50 Hz, (ii) to measure the phase difference between two voltages of the same frequency. When the X-frequency is 100 Hz and an unknown voltage V is joined to the Y-plates, a figure 8 is obtained on the screen. What is the frequency of V?

10 Draw a labelled diagram showing approximate potential values of the various electrodes in a cathode ray tube. On the diagram, add the power supply for the filament and the voltage supply for the electrodes.

11 Explain, by reference to electric fields, how an electron beam is focused on the screen of a cathode ray tube. Why is it inadvisable to have a stationary bright spot on the screen?

12 What is the purpose of X- and Y-shifts in a cathode ray tube? Explain, with a circuit diagram, how X- and Y-shifts are obtained.

Draw a block diagram of a cathode ray tube, showing the power supplies, the time-base circuit, and the signal circuit.

Draw a sketch of a time-base voltage variation, and explain how stationary waveforms are obtained.

13 Explain how a cathode ray oscilloscope can be used as (i) d.c. voltmeter, (ii) an a.c. voltmeter. Describe how the d.c. 'voltage sensitivity' of an oscilloscope can be measured.

Junction diode. Transistor

14 Explain the meanings of the following terms:
(i) valence electron, (ii) hole, (iii) impurity atom, (iv) thermal run-away. Why is silicon preferred to germanium in making semi-conductor devices?

15 What is p-type and n-type germanium or silicon? Starting with pure germanium, explain briefly how each type is made.

16 What is a *p-n junction diode*? Draw a sketch of its characteristic curve and explain its appearance from the movement of charge carriers. Describe, with diagram, how a bridge rectifier circuit works.

17 Draw a sketch of a *n-p-n* and a *p-n-p transistor*. Why is the *n-p-n* transistor preferred for high frequency circuits? Draw a diagram showing how voltage supplies are connected in the *common-base* arrangement for each transistor. Explain the action of the transistor in a common-base arrangement.

18 Draw a circuit diagram of a *common-emitter* arrangement, using a *n-p-n* transistor. Explain how current amplification is obtained in a common-emitter arrangement.

19 Draw sketches of the *input characteristics* $(I_b - V_b)$ and the *output characteristics* $(I_c - V_c)$ of a common-emitter transistor arrangement.

Account for the principal features of the two sets of curves by reference to transistor action.

What important parameters of the transistor can be calculated from these two graphs, and how are they obtained?

Fig. 50A.

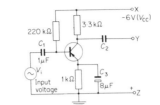

20 Fig. 50A shows a simple form of common-emitter transistor amplifier. (i) Is the transistor *p-n-p* or *n-p-n*? (ii) What is the purpose of the capacitor C_1, and why is it made fairly large? (iii) Which are the output terminals? (iv) Which capacitor in the circuit is used to prevent undesirable feedback? (v) Which resistor is used (a) as the load, (b) to stabilize the circuit for undesirable temperature rise, (c) to bias the base? (vi) Explain why the actual d.c. potential of the collector is negative relative to that of the base, and the actual d.c. potential of the emitter is positive relative to that of the base, when current flows.

Fig. 50B.

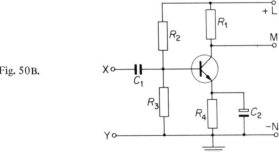

21 Fig. 50B shows a common-emitter amplifier circuit. Answer the same questions as in question 20.

Why is this circuit more widely used than the circuit in Fig. 50A?

51. PHOTOELECTRICITY. X-RAYS

(Where necessary, assume $e = 1.6 \times 10^{-19}$ C, $e/m_e = 1.8 \times 10^{11}$ C kg^{-1}, $c = 3.0 \times 10^8$ m s^{-1}, $h = 6.6 \times 10^{-34}$ J s)

Photoelectricity

1 On Einstein's photoelectric theory, the energy of photons in a monochromatic beam of frequency f is hf, where h is the Planck constant. Calculate the photon energy of (i) red light of frequency 0.4×10^{15} Hz, (ii) violet light of frequency 0.7×10^{15} Hz, (iii) ultra-violet light of frequency 1.0×10^{15} Hz.

2 Calculate the photon energy in monochromatic light of wavelength (i) 6.0×10^{-7} m (orange), (ii) 4.5×10^{-7} m (blue) in air.

3 The longest wavelength beyond which no photoelectrons are emitted from a metal surface is 7.0×10^{-7} m. Calculate the minimum energy required to liberate electrons from the metal surface, or 'work function' of the metal.

4 Using the metal of Question 3, what is the energy of photo-electrons emitted from its surface by light of wavelength 5.0×10^{-7} m?

5 An 'electron-volt (eV)' is a unit of energy equal to 1.6×10^{-19} J. Calculate the longest wavelength beyond which no photoelectrons are emitted from a caesium surface if its work function is 1.80 eV.

If the surface is illuminated by monochromatic light of wavelength 4.5×10^{-7} m, calculate the energy of the emitted photoelectrons.

What is the stopping potential which just prevents photoemission?

6 Repeat Question 5 for sodium, which has a work function of 2.46 eV.

7 State the experimental result in photoelectricity which could only be explained by Einstein's photon theory of light. Why is the wave theory unable to account for this result?

8 Describe a simple experiment to demonstrate the photoelectric effect. How is it deduced that negatively-charged, and not positively-charged, particles are emitted from a metal surface when suitably illuminated?

9 Write down two laws of photoelectricity. Describe, with circuit diagram, an experiment to determine the maximum kinetic energy of photoelectrons emitted by a metal illuminated by monochromatic light.

Monochromatic light is incident on the cathode of a vacuum photoelectric cell. Sketch the I-V graph obtained (i) for +ve and −ve V, (ii) when the cathode is replaced by a different metal.

10 Einstein's photoelectric equation is sometimes written as

$$hf = w_0 + \tfrac{1}{2}mv_m^2.$$

Explain the meaning of this equation.

In a photoelectric experiment, the retarding p.d. V is measured which just prevents photoemission. (i) Show that the relation $hc/\lambda - w_0 = eV$

applies, where λ is the incident wavelength. (ii) Explain how h can be determined from the results of the experiment.

11 The work function of a metal surface is 1.85 eV. Calculate (i) the maximum velocity of electrons emitted from the surface when illuminated by monochromatic light of wavelength 4.20×10^{-7} m, (ii) the stopping potential which just prevents photoemission in this case, (iii) the longest wavelength for which photoelectrons are emitted.

X-rays

12 What are *X-rays*? How do they differ from electrons? Draw a diagram of an X-ray tube and of the voltage supplies which operate it, and explain how X-rays are produced.

13 A p.d. of 40 kV is applied across an X-ray tube. Calculate the energy and speed with which electrons arrive at the anode, assuming the electrons have zero velocity at the cathode.

14 Calculate the quantum of energy (i) in joule and (ii) in electron-volts in monochromatic X-rays of wavelength 1.0×10^{-10} m. What evidence shows that the energy E in X-rays may be found from $E = hf$.

15 An X-ray tube is operated at 60 kV. What is the frequency of the most energetic radiation obtained? What is the minimum wavelength obtained? Calculate the new minimum wavelength if the voltage is changed to 30 kV.

16 An X-ray tube has a voltage of 40 kV. Show that the shortest wavelength obtained is about 3×10^{-11} m. Why are longer wavelengths obtained but not shorter wavelengths?

17 An X-ray tube produces a spectrum of one or more prominent lines? together with a background of continuous radiation having an abrupt minimum wavelength. Explain the reason for (i) the prominent lines, (ii) the minimum wavelength. What would be altered to change (iii) the sharp minimum wavelength value, (iv) the wavelengths of the prominent lines?

18 State *Bragg's law*. Calculate the wavelength of the first order diffraction image obtained by reflection at atomic planes of spacing 2.8×10^{-10} m in rocksalt if the glancing angle is then 6.2°. At what glancing angle is the second order diffraction image obtained?

19 The wavelength of a monochromatic X-ray beam is 1.0×10^{-10} m. Calculate the glancing angle obtained for the first order diffraction image by reflection at atomic planes of spacing (i) 2.0×10^{-10} m, (ii) 0.8×10^{-10} m.

20 When a monochromatic beam of X-rays is incident on a crystal it is reflected strongly in a particular direction making an angle θ with the crystal. (i) Draw a sketch in illustration. (ii) Write down the relation between θ and the incident wavelength λ. (iii) Draw a sketch showing how a monochromatic beam may be obtained from the spectrum of an X-ray tube by using a crystal, and write down the magnitude of the particular wavelength obtained.

21 What changes in the atom produce X-rays? Explain why the frequency of the characteristic wavelengths of elements is related to their atomic number. Draw a linear graph which illustrates the relation.

22 A second order diffraction image is obtained by reflection of X-rays at atomic planes of a crystal for a glancing angle of 11.4°. If the atomic spacing of the planes is 2.0×10^{-10} m, calculate the wavelength.

23 The minimum wavelength obtained from an X-ray tube is 0.4×10^{-10} m. Calculate the voltage at which the tube is operated. If this wavelength produces a first order diffraction image when the glancing angle on a crystal is 4.0°, calculate the atomic spacing of the reflecting planes.

52. ENERGY LEVELS IN THE ATOM

(Assume $c = 3.0 \times 10^8$ m s^{-1}, $e = 1.6 \times 10^{-19}$ C, $h = 6.6 \times 10^{-34}$ J s)

1 The first excitation potential of mercury atoms is 4.9 V. Calculate the wavelength of the emitted radiation if the excited atoms return to the ground state.

2 When sodium atoms are first excited, the emitted wavelength is 5.89×10^{-7} m. Show that this agrees with the first excitation potential observed, which was 2.1 V.

3 In a Franck-Hertz experiment, an *accelerating p.d.* is provided between the cathode and a grid, and a *retarding p.d.* between the grid and anode. Low pressure neon gas is contained in the tube. (i) What is the purpose of each p.d.? (ii) Why is the retarding p.d. a low value? (iii) Why is the neon gas at low pressure? (iv) If the anode current is relatively high, what type of collisions are made between the electrons and the neon molecules? (v) If the anode current falls at one stage as the accelerating p.d. is increased from zero, what type of collisions are now made by some electrons? (vi) Draw a rough sketch of the variation of anode current with accelerating p.d. from zero p.d. (vii) What conclusions may be drawn from the experiment?

4 The ground state of the hydrogen atom is −13.4 eV, and the next two energy levels are −3.34 eV and −1.5 eV respectively. A hydrogen atom is excited to the level −1.5 eV from the ground state and falls to the −3.34 eV level and then to the ground state. (i) Calculate the wavelength of the two radiations emitted. (ii) State the *ionization potential* of hydrogen and explain your answer.

5 Four prominent lines in the hydrogen visible spectrum have the following wavelengths:

4.10×10^{-7} m (violet); 4.34×10^{-7} m (blue);
4.86×10^{-7} m (green-blue); 6.56×10^{-7} m (red).

Each radiation is emitted by a fall in energy of the excited atom to a level −3.40 eV.

Calculate the energy level to which the atom is excited in each case. Show in a diagram, roughly to scale, the transitions of energy which produce the four radiations.

6 Define the terms *excitation potential, ionization potential, ground state* in connection with the atom.

7 Why is the hydrogen spectrum considered to help the concept of energy levels in atoms?

Electrons in hydrogen atoms can be excited to higher energy levels by absorbing energy from incident radiation of suitable wavelength. The ground state of hydrogen is −13.6 ev and the next energy level is −3.34 eV. Calculate the wavelength of radiation incident on hydrogen in the ground state which excites the electron to the next energy level. In which part of the electromagnetic spectrum does it occur? Why is hydrogen *transparent* to visible light?

53. RADIOACTIVITY. THE NUCLEUS. NUCLEAR ENERGY

Radioactivity

1 What are *α-particles, β-particles, γ-rays*? Draw a sketch showing their respective paths in a powerful magnetic field normal to their direction.

2 List in order (i) the penetrating power, (ii) the range in air at normal atmospheric pressure, and (iii) the ionizing power of α-, β-particles and γ-rays.

3 An electron, a proton and an α-particle have respective charge-mass ratios (specific charges) of 3680 : 2 : 1. If the three particles are accelerated from rest in a vacuum by the same p.d., what is the ratio of the respective velocities gained?

4 A particular form of Geiger-Müller tube has (i) a thin mica window, (ii) argon gas at low pressure, (iii) halogen gas mixed with the argon. Explain the purpose of these items.

Draw a sketch showing how the number of pulses per second varies when the applied voltage is increased from zero.

5 In an experiment to investigate the variation of the intensity I of γ-rays with the distance r from the source S, a GM tube is connected to a scaler and the distance d from S to the front of the GM tube is measured for various values of I.

(i) Draw a sketch of the arrangement. (ii) Why is a small source used? (iii) If C is the count rate at a distance d, what graph would be plotted to show how I varied with r? State the result.

6 'The half-life of Radon-220 is about 56 s.' What does this mean? If the count rate of this radioactive element is 600 per second at the start of observations, in what time would the count rate fall to 75 per second?

7 In an experiment to measure the half-life of a radioactive isotope, a GM tube was connected to a scaler and counts were taken every 30 seconds for 10-second intervals. Allowing for the background count, the following results were obtained:

Time from start and interval	0–10	40–50	80–90	120–130	160–170	200–210
Count	4260	2890	2025	1390	960	630

Plot the count rate C against the mean time t from the start, and hence deduce the half-life.

What other graph could be plotted to determine the half-life?

8 Describe, with experimental detail, how you would show (i) that a particular radioactive source X emits mainly β-particles and that another source Y emits mainly γ-rays, (ii) that radioactive decay is a random process.

9 (i) Define *half-life* $T_{1/2}$, *decay constant* λ. (ii) Draw a sketch to illustrate the decay of activity of a radioactive element with time, show on it the half-life period $T_{1/2}$, and write down the relation between the number N of undisintegrated atoms after a time t and the initial number N_0.

10 The half-life of a radioactive element is 24 days. Calculate the time taken for the activity to decay to (i) 0.2, (ii) 0.1 times its initial activity.

11 The half-life of radium is about 1600 years and its atomic mass is 226. Calculate (i) the percentage of undisintegrated atoms left in 100 years, (ii) the number of disintegrations per second in 1 milligram of radium, assuming one disintegration per atom. Avogadro constant = 6×10^{23} mol^{-1}.

12 Describe briefly an experiment to find the range of α-particles in air. 2×10^{12} atoms of a radioactive source X initially emits α-particles with a release of 4.0 mW of power. Each emission produces 5×10^{-13} J of energy. Calculate the initial rate of emission of α-particles and the half-life of X.

Nucleus. Nuclear Energy

13 Describe and explain briefly the experiment which led to the discovery of the nucleus by Rutherford.

14 (i) A nucleus of mass number 211 and atomic number 83 emits an α-particle followed by a β-particle (electron). What is the mass number and atomic number of the resultant nucleus?

(ii) A nucleus X emits a β-particle ($\beta-$); another nucleus Y emits a positron ($\beta+$). What nuclear change has occurred in each case?

15 A helium, nitrogen and uranium nuclei are denoted respectively by $^{4}_{2}\text{He}$, $^{14}_{7}\text{N}$, $^{238}_{92}\text{U}$. What are the number of electrons, protons and neutrons in each of the corresponding normal atoms?

16 In a nuclear collision, there is conservation of charge and of mass. Write down the nuclear equation when an α-particle ($^{4}_{2}\text{He}$) is incident on a

nitrogen nucleus ($^{14}_{7}$N) and a proton ($^{1}_{1}$H) is produced. Explain why this reaction indicates a "transmutation" of a nitrogen atom, and name the element of the new atom formed.

17 Einstein's mass-energy relation is '$E = mc^2$'. Explain the meaning of E, m, c. Estimate the energy released in joule by a mass change of 1 milligram.

18 The relative atomic mass of the carbon isotope ^{12}C is 12 g, and this is the mass of about 6×10^{23} carbon atoms. If the mass of a carbon atom is defined as 12 u, where u is the 'unified atomic mass unit', estimate the energy in joule equivalent to 1 u, using Einstein's relation. If 1 MeV (megelectron-volt) of energy $= 1.6 \times 10^{-13}$ J, calculate the magnitude of 1 u in terms of MeV.

19 Define the terms *nucleons, binding energy*. A neutron has a mass of 1.0087 u and a proton has a mass of 1.0073 u. The helium nucleus $^{4}_{2}$He, has a mass of 4.0015 u. Calculate the binding energy per nucleon of the helium nucleus in u.

20 A nuclear reaction is expressed by:

$$^{7}_{3}\text{Li} + ^{1}_{1}\text{H} \rightarrow ^{4}_{2}\text{He} + ^{4}_{2}\text{He} + Q.$$

(i) Write down in words the meaning of this reaction, if Q is the energy released; (ii) if the atomic masses of lithium, hydrogen and helium are 7.020, 1.008 and 4.003 u respectively, calculate Q in u.

21 What is *fission*? The uranium nucleus $^{235}_{92}$U undergoes fission by a neutron, $^{1}_{0}$n, according to the following reaction:

$$^{235}_{92}\text{U} + ^{1}_{0}\text{n} \rightarrow ^{141}_{56}\text{Ba} + ^{92}_{36}\text{Kr} + 3^{1}_{0}\text{n} + Q,$$

where Q is the energy released. (i) Why is a neutron capable of producing fission? (ii) Calculate Q in u if the atomic masses are $^{235}_{92}$U $= 235.1$ u, $^{141}_{56}$Ba $= 141.0$ u, $^{92}_{36}$Kr $= 91.9$ u, $^{1}_{0}$n $= 1.009$ u. (iii) Why can a *chain reaction* be obtained?

22 What are *isotopes*? How are they identified by a mass spectrometer? In a Bainbridge mass spectrometer, the magnesium ions ^{24}Mg$^+$ and ^{26}Mg^{2+} are deflected in circular paths by a uniform magnetic field. Find (i) the ratio of the 'specific charges' of the two ions, (ii) the radius of the path of the heavier ion if that of the lighter ion is 0.360 m.

23 Explain briefly how a *cloud chamber* works. Draw, and label, sketches illustrating the tracks in it of (i) α-particles from uranium, (ii) a nuclear collision. Why have some tracks perpendicular forks when α-particles pass through helium gas?

24 What is the difference between *fission* and *fusion*? In which kind of elements can these respective processes take place?

Draw a curve showing roughly how the *binding energy per nucleon* in the nucleus of different elements varies with the *mass number*. How is nuclear fission and nuclear fusion accounted for by reference to this curve?

25 The helium nucleus $^4_2$He may be produced by (i) the fusion of two deuterium nuclei $^2_1$H *or* by (ii) the fusion of two protons $^1_1$H and two neutrons $^1_0$n.

Calculate the energy produced in each case.

(Assume mass of helium = 4.0028 u, deuterium = 2.0142 u, proton = 1.0076 u, neutron = 1.0090 u and 1 u ≡ 931 MeV)

WORKED EXAMPLES ON ATOMIC PHYSICS

1 An electron beam passes undeflected between two plates 0.5 cm apart and having a p.d. of 150 volts when a perpendicular magnetic field of 5×10^{-3} T is applied. Find the velocity of the electrons, and the radius of the path travelled in the magnetic field alone. ($e/m = 1.76 \times 10^{11}$ C kg^{-1}.)

With the usual notation, since the beam is undeflected,

$$Bev = Ee$$

$$\therefore v = \frac{E}{B}$$

But $B = 5 \times 10^{-3}$ T, $E = 150/(0.5 \times 10^{-2})$ V m^{-1}

$$\therefore v = \frac{150}{5 \times 10^{-3} \times 0.5 \times 10^{-2}}$$

$$= 6 \times 10^6 \text{ m s}^{-1}$$

With the magnetic field alone,

$$Bev = \frac{mv^2}{r}$$

$$\therefore r = \frac{mv}{eB}$$

$$= \frac{1 \times 6 \times 10^6}{1.76 \times 10^{11} \times 5 \times 10^{-3}}$$

$$= 7 \times 10^{-3} \text{ m}$$

2 An oil-drop has a charge of $24e$, where e is 1.6×10^{-19} C, and is between two plates 4 mm apart. The drop falls under gravity with a velocity of 6×10^{-4} m s^{-1}, and a p.d. of 1600 V between the plates makes the drop rise with a steady velocity v. If the viscosity of air is 1.8×10^{-5} N s m^{-2} and the density of oil is 900 kg m^{-3}, calculate the radius of the drop and v. (Neglect the density of air.)

If a is the radius of the drop, then, with the usual notation, for fall under gravity,

$$mg = 6\pi\eta a v_1$$

or $$\frac{4}{3}\pi a^3 \rho g = 6\pi\eta a v_1$$

$$\therefore a = \left[\frac{9\eta v_1}{2\rho g}\right]^{1/2}$$

$$= \left[\frac{9 \times 1.8 \times 10^{-5} \times 6 \times 10^{-4}}{2 \times 900 \times 9.8}\right]^{1/2}$$

$$= 2.3 \times 10^{-6} \text{ m} \quad \ldots \ldots \ldots \ldots \quad \text{(i)}$$

With the field applied,

$$EQ - mg = 6\pi\eta a v$$

$$\therefore EQ = 6\pi\eta a(v + v_1), \text{ from above}$$

$$\therefore v + v_1 = \frac{EQ}{6\pi\eta a}$$

$$\therefore 6 \times 10^{-4} + v_1 = \frac{1600 \times 24 \times 1.6 \times 10^{-19}}{4 \times 10^{-3} \times 6\pi \times 1.8 \times 10^{-5} \times 2.3 \times 10^{-6}}$$

$$= 19.7 \times 10^{-4} \text{ m s}^{-1}$$

$$\therefore v_1 = 13.7 \times 10^{-4} \text{ m s}^{-1}$$

3 A metal has a work function of 2.0 eV and is illuminated by monochromatic light of wavelength 5.0×10^{-7} m. Calculate (i) the threshold wavelength, (ii) the maximum energy of the photoelectrons, (iii) the minimum retarding or "stopping" potential. ($1 \text{ eV} = 1.6 \times 10^{-19}$ J, $h = 6.6 \times 10^{-34}$ J s, $e = 1.6 \times 10^{-19}$ C, $c = 3 \times 10^8$ m s^{-1}.)

Quantum of energy in incident photons = $h\nu$.

(i) $\therefore$ minimum frequency of photons to liberate electrons is given by

$$h\nu_0 = 2.0 \text{ eV} = 2.0 \times 1.6 \times 10^{-19}$$

$$\therefore \frac{hc}{\lambda_0} = 2.0 \times 1.6 \times 10^{-19}$$

$$\therefore \lambda_0 = \frac{6.6 \times 10^{-34} \times 3 \times 10^8}{2.0 \times 1.6 \times 10^{-19}}$$

$$= 6.2 \times 10^{-7} \text{ m} = \text{threshold wavelength}$$

(ii) Maximum energy

$$= h\nu - \omega_0 \text{ (work function)}$$

$$= \frac{6.6 \times 10^{-34} \times 3 \times 10^8}{5.0 \times 10^{-7}} - 2 \times 1.6 \times 10^{-19}$$

$$= 0.8 \times 10^{-19} \text{ J}$$

(iii) Stopping potential, V, is given by
$$eV = 0.8 \times 10^{-19} \text{ J}$$
$$\therefore \quad V = \frac{0.8 \times 10^{-19}}{1.6 \times 10^{-19}} = 0.5 \text{ V}$$

4 The lithium nucleus is ^7_3Li. Calculate the binding energy in u if the proton has a mass of 1.0073 u, a neutron has a mass of 1.0087 u and the mass of the nucleus is 7.0180 u.

The nucleus has 3 protons and 4 neutrons

Mass of 3 protons = $3 \times 1.0073 = 3.0219$ u,
and
mass of 4 neutrons = $4 \times 1.0087 = 4.0348$ u

$\therefore$ total mass = 7.0567 u

$\therefore$ binding energy = mass of protons and neutrons − mass of nucleus
$$= 7.0567 - 7.0180$$
$$= 0.0387 \text{ u}.$$

5 The half-life of a radioactive element is 40 days. Calculate the time taken for the activity to decay to 30 per cent of its initial value.

If N_0 and N are the initial and final number of undisintegrated atoms,
$$N = N_0 e^{-\lambda t} \quad \ldots \ldots \ldots \text{(i)}$$
Now the half-life T is given by $2 = e^{\lambda T}$, or $e^{-\lambda} = 2^{-1/T}$.
$$\therefore \quad N = N_0 2^{-t/T} \quad \ldots \ldots \ldots \text{(ii)}$$
When $N/N_0 = 0.3$,
$$\therefore \quad 0.3 = 2^{-t/T} = \frac{3}{10}$$
$$\therefore \quad 2^{t/T} = \frac{10}{3}$$
$$\therefore \quad \frac{t}{T} \log_{10} 2 = \log_{10} 10 - \log_{10} 3$$
$$\therefore \quad \frac{t}{T} \times 0.3010 = 0.5229$$
$$\therefore \quad t = \frac{0.5229}{0.3010} T = \frac{0.523}{0.301} \times 40 \text{ days}$$
$$= 70 \text{ days}$$

6 The element in question 6 has an atomic mass of 220 and is monatomic. Find the number of emissions per second by a milligram of the substance after 30 days, assume one disintegration per atom. (Assume Avogadro constant is 6×10^{23} mol^{-1}.)

From $\quad N = N_0 e^{-\lambda t}$,

Differentiating,

$$\therefore \frac{dN}{dt} = -\lambda N_0 e^{-\lambda T} = -\lambda N_0 2^{-t/T}$$

But $\quad \lambda = \dfrac{0.693}{T} = \dfrac{0.693}{40 \times 24 \times 3600}, \quad t = 30$ days,

and $\quad N_0 =$ no. of atoms in 1 milligram $= \dfrac{1}{1000} \times \dfrac{1}{220} \times 6 \times 10^{23}$.

$$\therefore \frac{dN}{dt} = \text{no. of emissions per second}$$

$$= \frac{0.693 \times 6 \times 10^{23}}{40 \times 24 \times 3600 \times 1000 \times 220} \times 2^{-30/40} \text{ per second}$$

$$= 3.3 \times 10^{11} \text{ s}^{-1}$$

7 (i) Calculate the minimum wavelength obtained from an X-ray tube when the p.d. across it is 60 kV.

(ii) An X-ray beam of wavelength 0.8×10^{-10} m is incident on a crystal whose reflecting atomic planes have a spacing of 2.8×10^{-10} m. Calculate the glancing angle for the first order diffraction image.

(iii) From quantum theory, the maximum frequency ν_m is obtained when

$$h\nu_m = eV$$

$$\therefore \frac{hc}{\lambda_{min}} = eV$$

$$\therefore \lambda_{min} = \frac{hc}{eV} \quad \ldots \ldots \ldots \text{(i)}$$

But $h = 6.6 \times 10^{-34}$ J s, $c = 3 \times 10^8$ m s^{-1}, $e = 1.6 \times 10^{-19}$ C, $V = 60\,000$ V.

$$\therefore \lambda_{min} = \frac{6.6 \times 10^{-34} \times 3 \times 10^8}{1.6 \times 10^{-19} \times 60\,000} \text{ metres}$$

$$= 0.2 \times 10^{-10} \text{ m}$$

(iv) From Bragg's law, $\quad 2d \sin \theta = m\lambda$

$$\therefore 2 \times 2.8 \times 10^{-10} \sin \theta = 0.8 \times 10^{-10}$$

$$\therefore \sin \theta = \frac{0.8}{2 \times 2.8} = 0.143$$

$$\therefore \theta = 8.2°$$

54. MULTIPLE CHOICE QUESTIONS – ELECTRICITY AND ATOMIC PHYSICS

For each of the questions **1–25**, *choose one statement from* **A** *to* **E** *which is the most appropriate.*

1. In electrical units,
   A the watt is the unit of energy.
   B one ampere = one coulomb x one second.
   C one ohm = one ampere/one volt.
   D one volt = one joule/one coulomb.
   E one watt = one volt x one coulomb.
2. When a current of 0.2 A flows through a cell of e.m.f. 1.1 V and internal resistance 2 Ω in a direction against the e.m.f., the terminal p.d. is:
   A 1.5 V   B 1.1 V   C 0.7 V   D 1.0 V   E 1.2 V
3. The relative atomic mass of silver is 108 and its valency is 1. If the Faraday constant is 96 500 C mol^{-1}, the current flowing when 1.08 g of silver is deposited in 1000 seconds is:
   A 0.108 A   B 0.4825 A   C 1.08 A   D 0.965 A   E 9.65 A
4. A cell of e.m.f. 10 V and internal resistance 5 Ω is used to supply power to an external resistance R. When maximum power is developed in R, the current flowing is:
   A 2 A   B 1.8 A   C 1.2 A   D 1.1 A   E 1.0 A
5. In SI units:
   A $\mu_0 = 1$ H m^{-1} and $\epsilon_0 = 8.85 \times 10^{-12}$ F m^{-1}
   B $\mu_0 = 4\pi \times 10^{-7}$ H m^{-1} and $\epsilon_0 = 1$
   C $\mu_0 = 4\pi \times 10^{-7}$ H m^{-1} and ϵ_0 may be found from $3 \times 10^8 = 1/\sqrt{\mu_0 \epsilon_0}$ (approx.)
   D $\mu_0 = 4\pi \times 10^{-7}$ F m^{-1} and $\epsilon_0 = 8.85 \times 10^{-12}$ H m^{-1}
   E 8.85×10^{-12} F m^{-1} is an exact value for ϵ_0 and $4\pi \times 10^{-7}$ H m^{-1} is approximate for μ_0.
6. The resistivities of two wires X and Y are in the ratio 1 : 2, their lengths are in the ratio 1 : 2, and their diameters are in the ratio 1 : 2. The ratio of the resistances of X and Y is then:
   A 1 : 2   B 1 : 1   C 2 : 1   D 4 : 1   E 8 : 1
7. A current of 5 A flows downwards in a long straight vertical conductor and the earth's horizontal flux density is 2×10^{-5} T. Then the neutral point is:
   A due north and 10 cm from the wire,
   B due east and 10 cm from the wire,
   C due east and 5 cm from the wire,
   D due west and 5 cm from the wire,
   E due south and 10 cm from the wire.
8. Two parallel straight conductors at a distance r apart and carrying currents in the same direction
   A repel each other with a force proportional to r,
   B attract each other with a force inversely-proportional to r^2,

C repel each other with a force inversely-proportional to r,
D attract each other with a force inversely-proportional to r,
E have a magnetic field with maximum flux density midway between them.

9 In measurement of alternating current or voltage:
A a hot-wire ammeter measures peak values directly
B a copper voltameter can be used to measure r.m.s. values
C an oscilloscope measures r.m.s. values directly
D a moving-coil meter can never be adapted to measure a.c.
E a moving-coil meter plus rectifier can measure r.m.s. values of a.c.

10 In a moving-coil meter or galvanometer:
A the sensitivity is the current per unit deflection
B the torque due to a current depends on the strength of the springs or torsion wire
C the sensitivity may be expressed in millimetres per microamp
D the torque due to a current does not depend on the radial field present
E the linearity of the scale is due to the control exerted by the springs or torsion wire

11 With a step-up commercial iron-cored transformer:
A the flux linkage of the primary and secondary coils is always high
B radio-frequency voltages cannot be stepped up owing to high energy losses
C losses of energy occur only in the iron
D removal of the iron makes no difference to the transformer action
E the current in the secondary is higher than in the primary

12 A voltmeter is connected across the terminals of a d.c. motor joined to a suitable battery. When the motor is used to rotate a machine X and current flows:
A the voltmeter reads the e.m.f. of the battery
B the voltmeter reads the back e.m.f. in the motor
C the voltmeter reading is a measure of the power supplied to X
D the voltmeter reads the energy per coulomb supplied to X
E the voltmeter reading is inversely-proportional to the speed of X

13 A potentiometer with a connected circuit are shown in Fig. 54A. State what happens to the balance point O when the following operations are carried out: (a) Increase R_3 to a very high value, (b) decrease R_1 to zero, (c) decrease R_3 to zero, (d) increase R_2.

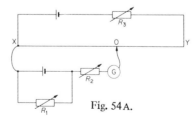

Fig. 54A.

Choose your answers from:
A O stays where it is

ATOMIC PHYSICS

B O moves to the left but not as far as X
C O moves to the right, being at Y before the end of the operation
D O moves to the left, being at X at the end of the operation
E O moves to the right, being at Y at the end of the operation

14 An electric fire is found to give sparks when its case is touched onto other earthed appliances, although it does not give a shock. The most likely error is:
A The earth wire has not been connected
B the live and neutral wires have been interchanged
C the earth and neutral wires have been interchanged
D the live and the earth wires have been interchanged
E the neutral wire has not been connected.

15 An a.c. circuit contains a total resistance of 100 Ω and a total reactance of 100 Ω. When a current of 10 A (r.m.s.) flows, the power absorbed is:
A 20 kW **B** $10\sqrt{2}$ kW **C** 100 kW **D** 10 kW **E** 2 kW

16 Three photographs, X, Y, Z, contain the following geometrical patterns: X-concentric circles, Y — parabolic arcs, Z—circular and spiralling paths.
A X may be due to electrons incident on a thin film of graphite
B X may be due to X-rays incident on a thin film
C Y may be due to electrons entering a perpendicular magnetic field
D Z may be due to electrons entering the electric field between parallel plates
E Y may be due to electrons entering a magnetic field parallel to its velocity

17 In a fine beam tube, electrons are accelerated by an anode voltage and then enter a perpendicular magnetic field due to direct current in two Helmholtz coils where they move in a circular path. Then:
A the radius is unaffected if an a.c. 'ripple' is superimposed on the direct
B the radius decreases when the anode voltage increases
C the radius is unaffected if an a.c. 'ripple' is superimposed on the direct current in the coils
D the radius is affected if an alternating voltage 'ripple' is superimposed on the anode voltage
E the radius is unaffected when alternating current flows in the coils in place of direct current

Fig. 54B.

PQ is a movable conducting roller on a fixed horizontal conducting frame JKLM connected to a battery. Fig. 54B. State what happens when the apparatus is placed in a uniform magnetic field: (*a*) acting vertically downwards, (*b*) acting in the direction LM, (*c*) acting in the direction PQ.

Choose answers from:
A roller moves to left,
B roller moves to right,
C roller tries to move down,
D nothing,
E roller bobs up and down.

19 In Fig. 54B, the roller PQ is 2 m and is moved with a speed of 2 m s^{-1}. The e.m.f. of the battery is 2 V and it has internal resistance 2 Ω, The resistance of PQ and the frame JKLM may be ignored. Calculate the current in the following cases:
(*a*) Field of 0.25 T vertically down and wire moves to the left
(*b*) Field of 0.25 T vertically down and wire moves to the right
(*c*) Field of 0.25 T in the direction PQ and wire moves to right
(*d*) Field of 0.5 T vertically down and the wire moves to the right
Choose answers from: **A** 1.0 A, **B** 2.0 A, **C** 1.5 A, **D** 0, **E** 0.5 A.

20 In a Geiger-Müller tube, a trace of inert gas is added to a halogen
A to counteract chemical action by the halogen gas,
B to quench the discharge,
C to act as a catalyst in the electrical action,
D to absorb the radioactive particle or radiation,
E to initiate the discharge.

21 To demonstrate the random nature of nuclear disintegration,
A a source of constant activity should be used,
B a source of half-life about ten hours should be used,
C a source of half-life about ten seconds should be used,
D a weak source such as that providing 'background' may be used,
E the distance of the source from the detector should be varied.

22 In an experiment to show that γ-rays obey an inverse-square law of intensity I with a distance r, it is best to plot values of $I^{-1/2}$ against d, where d is the distance from the source to a fixed point on the detector, rather than I against $1/d^2$. This is because
A it is difficult to measure r exactly,
B d is easier to measure than r,
C $I^{-1/2}$ is less than I and hence less error is obtained,
D d is easier to plot than $1/d^2$,
E γ-rays travel in straight lines over a short distance d.

23 Using a radioactive source of constant activity, the count rate diminishes from 1024 to 128 in 3 minutes. In 4 min, the count rate diminishes further to:
A 107 **B** 82 **C** 64 **D** 56 **E** 32

24 When a radioactive sample disintegrates, the rate of disintegration at any instant is proportional to
A the original mass of the sample,
B the mass of all the new products of disintegration,
C the number of undisintegrated atoms,
D the atomic number of the radioactive elements,
E the external atmosphere pressure.

ATOMIC PHYSICS

Instructions

In the following *Assertion–Reason* questions, answer:
A if the assertion, the reason, and the explanation are all true,
B if the assertion and reason are both true but the explanation is false in the sense that it does NOT fit the assertion.
C if the assertion is true and the reason is false,
D if the assertion is false and the reason is true,
E if the assertion and reason are both false.

	Assertion	*Reason*
25	Outside a straight wire, $B = \mu_0 I/2\pi r$	Integration of $\mu_0 I.\delta l \sin \theta /r^2$
26	E.m.f. of a cell is the total energy per coulomb	P.d. is the energy per coulomb
27	Heat in a uniform wire at constant p.d. is inversely proportional to its length	Pure metals have a lower resistance than alloys
28	An induced current can be obtained in a coil even when its ends are not connected	Induced e.m.f. $\propto$ rate of change of flux
29	Stationary electric charges deflect a magnetic needle	The electric and magnetic fields interact with each other
30	The electric field inside a hollow sphere at 50 kV is zero	The capacitance of a sphere is proportional to its radius
31	The reactance of a capacitor decreases as the frequency increases	The capacitance increases as the frequency increases
32	A junction diode or diode valve may sometimes amplify alternating voltages	The diodes act as rectifiers of a.c. voltage
33	A metal disc, suspended above a spinning magnet, will rotate in the same direction as the magnet	This follows from Lenz's law
34	Wires carrying parallel currents in the same direction will attract each other	The magnetic field at one wire X due to current in the other wire is parallel to X
35	The masses of copper and silver deposited in voltameters in series are proportional to their relative atomic masses	Copper and silver ions carry equal charges
36	The self inductance of an air-cored coil of given length and area is proportional to the number of turns	L = flux per unit current linking the coil

#	Statement	Reason
37	In an a.c. series L, C, R circuit, resonance is obtained only at one frequency	$I = V/Z$ at resonance
38	An electron beam, entering the magnetic field between Helmholtz coils perpendicular to its direction, is deflected in a circular arc	The coils are perpendicular to each other
39	X-rays have wavelengths of the order of 10^{-10} metre	They are produced by energy changes of electrons close to the nucleus
40	Photoelectrons of high energy will be rejected by infra-red radiation	Energy in photon is $h\nu$
41	A uranium nucleus can produce energy by fission	The bombarding neutron repels the nucleus violently
42	The charge-mass ratio for all the nuclei of chlorine atoms is the same	All chlorine atoms are alike
43	α-particles produce straight tracks in cloud chambers which can be photographed	α-particles are charged particles
44	An electron beam, entering a uniform electric field perpendicular to its direction, is deflected in a parabolic path	The x and y distances travelled by an electron are related by $x = vt$, $y = \frac{1}{2}at^2$, where $a =$ acceleration $= Ee/m$ and v is the initial velocity of the electron
45	The charge-mass ratio for a heavy hydrogen, $^2_1$H, nucleus is the same as that for a helium, $^4_2$He, nucleus	The mass of a helium nucleus is twice that of a hydrogen nucleus
46	A powerful electric lamp will always produce photoelectrons when illuminating a metal plate	Energetic light emits photoelectrons
47	Neon is chemically inert	The mass number is 20
48	In positive-ray experiments, hydrogen ions produce the parabola of greatest dimensions	Hydrogens ions have the smallest charge-mass ratio
49	γ-rays produce intense ionization in a cloud chamber	γ-rays produce straight thick tracks
50	In Millikan's oil drop experiment, the terminal velocity is measured under gravity alone	To provide the radius of the drop
51	Isotopes are atoms with equal masses	The nuclear charges on isotopes are equal
52	Electrons are produced in discharge tubes at a low pres-	Air conducts electricity at these pressures

sure of 0.01 mmHg

53 If green light produces photoelectrons from a metal surface, then violet light will always produce photoelectrons	The amplitude of vibration in violet light is greater than in green light
54 In Thomson's e/m experiment, only a magnetic field is applied	e/m can be found directly from $Bev = mvr$ since v cancels

55. MULTIPLE CHOICE QUESTIONS – GRAPHS

Optics

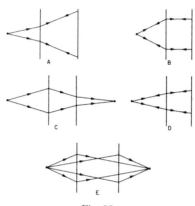

Fig. 55A.

Each of the diagrams A–E represents a method of measuring the following optical constants. Copy each diagram, identify the method, and draw in the diagram the optical components concerned and label them.

1. Converging lens focal length by plane mirror method.
2. Displacement method of measuring focal length of converging lens.
3. Concave mirror method of measuring focal length of diverging lens.
4. Radius of curvature of converging lens by Boys' method.
5. Converging lens method of measuring focal length of diverging lens.

Electricity

In questions 6–25, the first-named quantity is plotted along the y (vertical)-axis and the second along the x (horizontal)-axis. Name one graph from A to O which most likely represents the variation in each question, copy the curve, and label the axes. Note that one graph in A to O may be used more than once in the answers to questions 6–25.

6. Deflection (θ) in hot-wire ammeter against current (I).
7. Reactance (X_L) of a pure inductor against frequency (f).

8 Energy (W) in a photon against wavelength (λ).
9 Terminal p.d. of cell (V) against current (I) from cell.
10 Current (I) against a.c. frequency (f) for L, C, R series circuit (V constant).
11 Current (I) against time (t) for L-C energy exchanges with resistance present.
12 Anode current (I_a) against anode voltage (V_a) for a triode (V_g constant).
13 Current (I) against p.d. (V) for dilute sulphuric acid with platinum electrodes.
14 $\sqrt{\text{frequency}}$ against Z (atomic number) for X-ray K-lines.
15 Resistance (R) of pure metal against diameter (d) (given material and length).

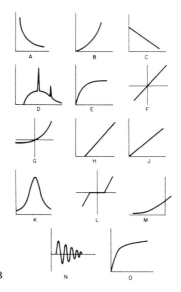

Fig. 55B

16 Current (I) against p.d. (V) for a junction diode.
17 Reactance (X_C) of capacitor against frequency (f).
18 Flux density (B) against $1/r$ for straight wire.
19 Charge (Q) against time (t) for charging a capacitor through high resistance.
20 Common-emitter transistor characteristic, I_C against V_{CE}, I_B constant.
21 Impedance (Z) against frequency (f) for C–R in series (a.c.).
22 Intensity (I) against wavelength (λ) for X-ray tube.
23 Triode mutual characteristic, I_a against V_g, V_a constant.
24 Impedance (Z) against frequency (f) for L–R in series (a.c.).
25 Current (I) against p.d. (V) for copper at constant temperature.
26 The graph in Fig. 55C shows the angle θ against time t when the coil X is rotated in a uniform horizontal magnetic field B.

Fig. 55C.

Which of the lettered graphs A to E shown in Fig. 55D most suitably describes the variation with time of the following quantities:
(a) the flux through the coil,
(b) the e.m.f. induced in the coil,
(c) the power dissipated in a resistor connected to the coil,
(d) the current in a capacitor connected to the coil,
(e) the current in an inductor connected to the coil,
(f) the current in a resistor connected to the coil.

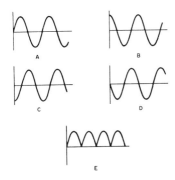

Fig. 55D.

27 The following graphs (Fig. 55E) are concerned with photo-electric emission from a potassium photocell.

Fig. 55E.

Label the axes (a) to (e), choosing from the quantities listed below:
A Photoelectric current, B electron velocity, C electron energy,
D p.d. across photocell, E frequency of light.

Electrostatics

Name one graph in A to F (Fig. 55F) which most nearly represents each of the variations 28–34.

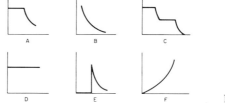

Fig. 55F.

28 Variation of potential V with distance r from a point charge.
29 Variation of potential V with distance r from the centre, inside a hollow charged conductor.
30 Variation of energy W of capacitor with voltage V.
31 Variation of potential V with distance r from the centre of a hollow charged conductor.
32 Variation of intensity E with distance r from the centre of a charged hollow conductor.
33 Variation of potential V from the centre of a hollow charged sphere with distance r from the centre, when an earth conductor is situated near it.
34 Variation of intensity E with distance r from a point charge.

56. MISCELLANEOUS MULTIPLE-CHOICE QUESTIONS

INSTRUCTIONS

In each of the questions which follow, four statements, (i)–(iv) are made.
    Answer **A** if only (i), (ii) and (iii) are correct.
    Answer **B** if only (i) and (iii) are correct.
    Answer **C** if only (ii) and (iv) are correct.
    Answer **D** if only (iv) is correct.
    Answer **E** if none is correct.

1 In an experiment to measure Young's modulus, two similar wires, loads from 0–5 kg, a vernier scale, a metre rule, and a micrometer gauge are used. In this case:
  (i) the initial length of wire to the load is measured by the metre rule and the gauge is used to determine the diameter.
  (ii) the two wires are used to eliminate error due to sag and the initial length is measured to the vernier.
  (iii) the loads are added 1 kg at a time and the average extension per kg is found by subtracting the first and last reading and dividing by five.
  (iv) the loads are added 1 kg at a time and the average extension for 3 kg is found by subtracting the readings for 0–3, 1–4, 2–5 kg and dividing by 3.

2 A load of 40 N is attached to a wire of length 4 metres and cross-sectional area 2×10^{-6} m^2, and produces an extension of 0.5 mm. In this case:

ATOMIC PHYSICS 139

(i) The strain is 1.25×10^{-4} and the energy stored is 10^{-2} J.
(ii) The strain is 2×10^7 and the molecules are further apart than normal.
(iii) The stress is 2×10^7 N m^{-2} and the energy in the wire is due to work done in separating the molecules.
(iv) Young modulus is 1.6×10^9 N m^{-2} and the molecules in the wire are closer than normal.

3 If a biconvex glass lens is fully immersed in water, then:
(i) the focal length is less than in air.
(ii) the lens acts as a diverging lens.
(iii) only real images are now obtained.
(iv) the focal length is longer than in air.

4 In temperature scales:
(i) on the thermodynamic scale, the upper fixed point is the triple point of water and the lower fixed point is absolute zero.
(ii) the platinum resistance thermometer scale utilises the relation $R_t = R_0(1 + \alpha t + \beta t^2)$ where t is the temperature on the gas thermometer scale.
(iii) the gas thermometer scale is used as a standard temperature scale.
(iv) the thermoelectric thermometer is not used at low temperatures.

5 An ideal gas:
(i) has molecules with low mutual attraction.
(ii) only obeys Boyle's law.
(iii) has a low critical temperature.
(iv) cools down on expansion into a vacuum.

6 In thermal conduction:
(i) metals are good conductors owing to the presence of free electrons.
(ii) mercury is a better conductor than copper.
(iii) radial flow of heat occurs across pipes.
(iv) thin glass is a better insulator than thick glass.

7 A black body radiator:
(i) emits only wavelengths in the infra-red region.
(ii) may be made practically by using a hole in a large closed can.
(iii) emits total energy proportional to the fifth power of the absolute temperature.
(iv) has maximum energy in a narrow band of wavelengths whose mean value is inversely-proportional to T, the absolute temperature.

8 Light is classified as a transverse wave on account of the phenomenon of:
(i) refraction.
(ii) interference.
(iii) diffraction.
(iv) polarization.

9 A galilean telescope has an objective of $f = 40$ cm, and in normal adjustment its angular magnification is 5. The separation of the lenses is then:
(i) 32 cm and the final image is erect.

(ii) 35 cm and the final image is erect.
(iii) 32 cm and the eyepiece has a focal length of 8 cm.
(iv) 45 cm and the eyepiece is a diverging lens.

10 When the earth moves in its orbit round the sun:
(i) its velocity increases as it approaches the sun.
(ii) its angular momentum about the sun is constant.
(iii) it has an elliptical path.
(iv) its total energy increases near the sun.

11 An elastic collision differs from an inelastic collision because:
(i) linear momentum is not conserved in an elastic collision.
(ii) an elastic collision does not occur between particles such as electrons and atoms.
(iii) the kinetic energy after an elastic collision is less than before.
(iv) the particles always coalesce after collision.

12 A mass M is attached to the end of a light spring S on a smooth table and the other end of S is fixed. When M is pulled slightly to extend the spring and then released:
(i) M and S vibrate in phase.
(ii) M has maximum kinetic energy when S has minimum potential energy.
(iii) the total energy of M and S is constant.
(iv) S has minimum potential energy when fully extended.

13
(i) Density is a vector quantity.
(ii) angular momentum is a scalar quantity.
(iii) work is a vector quantity.
(iv) pressure is a scalar.

14 When a light thin glass ring of radius r is lowered gently on to the surface of clean water of surface tension γ:
(i) the downward force on the ring is $4\pi r\gamma$
(ii) the vertical force on the ring is $2\pi r\gamma$
(iii) the angle of contact of the liquid with the ring is zero.
(iv) the surface tension force on the ring is $\pi r^2 \gamma$.

15 If a glass capillary tube is placed in water, which then rises to a height h above the outside level:
(i) the inside of the tube was wetted first.
(ii) h is inversely-proportional to the radius of the tube.
(iii) the angle of contact is zero only when the tube and water are both clean.
(iv) h is the height between the meniscus and the outside water level in contact with the tube.

16 A tuning fork and a violin, both sounding the same note:
(i) produce a longitudinal and a transverse sound wave respectively.
(ii) produce waves of the same amplitude.
(iii) produce waves of the same waveform.
(iv) produce waves of different waveform.

17 When a closed pipe of length l is sounding its overtone:
(i) The bottom and top of the pipe are both nodes of displacement.
(ii) The wavelength of the wave in the pipe is $4l/3$.

(iii) The first overtone is $2f_0$, where f_0 is the fundamental frequency.
(iv) The bottom of the pipe is an antinode of pressure and the top is a node of pressure.

18 When a sonometer wire is plucked in the middle:
(i) a thin wire produces a note of higher pitch than a thick wire of the same material.
(ii) doubling the tension produces a note of twice the original frequency.
(iii) decreasing the tension to one-quarter produces a note of half the original frequency.
(iv) the first overtone has a frequency $2f_0$, where f_0 is the fundamental frequency.

19 The speed of sound in a medium:
(i) depends on the density of the medium.
(ii) may be faster than light.
(iii) depends on the magnitude of the bulk modulus.
(iv) diminishes with pressure decrease in the case of air.

20 When a negatively-charged rod is brought near to the cap of a gold-leaf electroscope whose case is earthed:
(i) the leaf has an induced negative charge.
(ii) there is a p.d. between the cap and case.
(iii) the cap has a lower potential than the case.
(iv) the leaf has an induced positive charge.

21 When an air parallel-plate capacitor is connected to a battery, the energy between the plates increases when the air is replaced by a slab of mica because:
(i) the p.d. between the plates then increases.
(ii) the capacitance is reduced.
(iii) the electric intensity between the plates increases.
(iv) current flows from the battery thus increasing the charge.

22 Metal conductors can be distinguished from semiconductors because:
(i) metals have charge carriers of one sign only.
(ii) Ohm's law is not obeyed by a pure semiconductor.
(iii) metals increase in resistance with temperature rise and semiconductors decrease in resistance.
(iv) metal alloys can be used for standard resistors whereas impure semi-conductors have no useful applications.

23 When a resistance is measured by a metre bridge and by a Post Office box:
(i) the balance point in each case depends on the battery voltage used.
(ii) the accuracy in each case is the same.
(iii) there is an 'end-correction' in the cast of the Post Office box.
(iv) a galvanometer protective resistor is not needed for the metre bridge.

24 When a small motor is joined in series with a suitable light bulb and battery, and the current is switched on:
(i) the light is bright at first and then becomes dim.
(ii) the light has the same brightness throughout.

(iii) the speed of the motor is proportional to its back e.m.f.
(iv) the filament resistance decreases as the current increases.

25 When a coil of area A and N turns is rotated with constant angular velocity ω in a field of flux density B perpendicular to its axis:
   (i) the induced e.m.f. is proportional to ω^2.
   (ii) the maximum e.m.f. is given by NAB.
   (iii) the e.m.f. is zero at an instant when no flux links the coil.
   (iv) the e.m.f. at any instant is given by $NAB \sin \omega t$.

26 In the magnetic fields B due to current-carrying conductors:
   (i) $B = \mu_0 I/2\pi r$ outside a long straight conductor.
   (ii) $B = \mu_0 NI/2r$ in the middle of a flat circular coil.
   (iii) $B \propto 1/r^2$ according to the Biot-Savart law.
   (iv) $B = \mu_0 NI/2\pi l$ in the middle of a long solenoid of length l and N turns.

27 When an alternating current flows in a hot-wire ammeter and a moving-iron repulsion meter:
   (i) only the hot-wire ammeter has a deflection in it.
   (ii) the moving-iron meter records the peak value of the alternating current.
   (iii) one records the r.m.s. value of alternating current and the other the peak value.
   (iv) both record the r.m.s. value of alternating current.

28 When a water voltameter and silver voltameter are in series:
   (i) oxygen and silver are obtained at the cathodes.
   (ii) about 96 500 coulombs deposit 2 g of hydrogen and 108 g of silver.
   (iii) a hydrogen and oxygen ion each carry a charge numerically equal to e.
   (iv) the same number of hydrogen and silver ions are deposited in the same time.

29 In radioactivity:
   (i) β-particles cannot be detected in a cloud chamber.
   (ii) α-particles can be deflected by powerful magnetic field.
   (iii) γ-rays can never be emitted simultaneously with α-particles.
   (iv) β-particles can be deflected by an electric field.

30 In experiments on the electron by Thomson (e/m) and by Millikan (e):
   (i) the velocity of the electrons was needed to determine e/m.
   (ii) parallel electric and magnetic fields were applied in measuring e/m.
   (iii) the terminal velocity in air of an oil drop was measured in the determination of e.
   (iv) the relation $F = Bev$ was needed in the determination of e.

9
Answers

MECHANICS

EXERCISE 1 (p. 1)

1. 600 m
2. (i) 5 m; (ii) 7 m s^{-1}; (iii) 2 s
3. (i) 0.5 m s^{-2}; (ii) 2 m s^{-2}; 40 s, 10 s
4. 20 m, 40 m
5. (i) 2.5 J, 0.5 N s; (ii) 2×10^5 J, 2×10^4 N s
6. (i) 200 N s; (ii) 1000 J
7. (i) 5 J, 0; (ii) 2.5 J, 2.5 J
8. 0.45 N, 0.075 m s^{-1}
9. 40 000 J, 13 860 J
10. (i) 4×10^4 J; (ii) 16 m s^{-1}; (iii) 8×10^3 J
11. 4×10^5 N; 3.5×10^5 N
12. (i) 582 N; (ii) 618 N
13. 0.06 m s^{-1}
14. Law applies to both
15. $u = 5\sqrt{3}$ m s^{-1}, $v = 5$ m s^{-1}; yes
16. 4.5×10^{-19} J, 4.5×10^{-18} N
17. 8×10^{-24} N s. (i) 6400 N; (ii) 6.4×10^5 N m^{-2}
18. (i) 1000 m s^{-1}; (ii) 100 : 1
19. 9×10^{-12} N
20. LT^{-1}, LT^{-2}, ML^{2}T^{-2}, MLT^{-1}; all vectors except energy
25. 90°

EXERCISE 2 (p. 3)

1. (i) 0.2 m s^{-1}; (ii) 0.4 m s^{-2}; (iii) 10^{-3} J; (iv) 3.1 s
2. (i) 12.6 m s^{-1}; (ii) 400 m s^{-2}; (iii) 0.8 J; (iv) 0.2 s
3. 3617 km h^{-1}
4. 10 rad s^{-1}; bottom
5. 32°
6. (i) 6250 N; (ii) 11 790 N
7. 1.7 m s^{-1}
8. (i) 30 m s^{-1}; (ii) 195 N; (iii) 199 N
9. (i) 8.5×10^{-12} N; (ii) 2.1×10^{-12} J
10. (i) 2.4×10^6 m s^{-1}; (ii) 2.5×10^{-18} J
11. (i) 0; (ii) $2r\omega$; (iii) 0; (iv) $2mr\omega$
12. 6.36×10^{-6} m
13. 6250 s
14. 5009 s
15. 6.0×10^{24} kg
17. 9.85 m s^{-2}
18. 5500 kg m^{-3}
19. 3.7 m s^{-2}
20. (i) 8×10^3 m s^{-1}
21. 3.6×10^5 m
23. 315 : 1
24. (ii) 11.2 km s^{-1}

EXERCISE 3 (p. 5)

1. C, E
2. (i) 0.13 m s^{-1}, 0; (ii) 0.12 m s^{-2}, 0.1 m s^{-2}; (iii) 7.9×10^{-5} J
3. (i) 0.04 m s^{-1}; (ii) 0.016 N
4. 0.13 m s^{-1}, 0, 4.7×10^{-4} J
5. (i) 1.0 m; (ii) 0.06 m
7. 0.501 N
8. 200 N m^{-1}. (i) 0.31 s; (ii) 4×10^{-4} J; (iii) 4×10^{-4} J
9. 1.7×10^{12} Hz
10. (i) 0.2 s; (ii) 0.16 m s^{-1}, 1.2×10^{-4} J; (iii) 0.15 m s^{-2}, 0.116 N; 2×10^{-4} J

11. 2.3 Hz
12. 0.9 s
13. (i) p.e. = $\frac{1}{2}ka^2$, k.e. = 0;
    (ii) p.e. = 0, k.e. = $\frac{1}{2}ka^2$

EXERCISE 4 (p. 7)

1. (1) M.I.; (2) $I \times d\omega/dt$; (3) A.M.;
   (4) $I\omega$; (5) angle; (6) $\frac{1}{2}I\omega^2$
   (7) power; (8) power;
   (9) $I\omega \, d\omega/dt$
2. (i) 49.3 J; (ii) 3.1 kg m^2 s^{-1};
   (iii) 1.6 rev, 0.6 s
3. 1.1×10^{-34} kg m^2 s^{-1}
4. 2.7×10^{40} kg m^2 s^{-1}
5. (i) 0.63 kg m^2 s^{-1}; (ii) 1 rev s^{-1}
6. (i) 25 s; (ii) 12.6 J;
   (iii) 100.5 rev
7. (i) 0.12 kg m^2 s^{-1};
   (ii) 7.1 rad s^{-1}; (iii) 4.4 rad s^{-1}
8. (i) 8×10^{-2} N m; (ii) 0.8 rad s^{-2};
   (iii) 1.6 rad s^{-1};
   (iv) 12.8×10^{-2} J
10. 1.8 rad s^{-1}
11. 4 s
12. (i) 2.2 s; (ii) 1.9 s
13. 1 rad s^{-2}, 5 rad s^{-1}
14. 3.2 s
15. (i) 2 s; (ii) 2.4 s

EXERCISE 5 (p. 9)

1. 40 N
2. 58 N, 208 N
3. 82 N, 82 N
4. 100 N, 0.58
5. (i) 0.53 N; (ii) 0.78 N
6. (i) 180 N; (ii) 208 N
7. (ii) 160 N m; 50 N m
8. 0.04 rad
9. 6.3 cm
10. (i) 5.3 cm; (ii) 4.5 cm
12. (i) 0.9 : 1; (ii) 0.3 : 1
13. (i) 0.118 N; (ii) 0.094 N
14. (i) 1.3 N; (ii) 1.2 N

15. 2100 kg m^{-3}
16. (i) 0.12 m s^{-1} (ii) 6.75 N m^{-2}
17. 2 m s^{-1}, 6 kg s^{-1}
18. 5.55×10^4 Pa
19. 2 kg s^{-1}
20. 600 m s^{-1}, 60°
21. (i) 8000 J, 25.5 rev (ii) 2.5 m s^{-2}
23. (i) 8×10^{-3} J (ii) 0.28 m s^{-1}
    (iii) 2 m s^{-2}

PROPERTIES OF MATTER

EXERCISE 6 (p. 12)

1. 8.0×10^{28}
2. 2.6×10^{-10} m
3. 3.2×10^{-10} m
6. $(9a/b)^{1/8}$

EXERCISE 7 (p. 13)

1. (ii) ML^{-1} T^{-1}, nil, ML^{-1} T^{-2}
2. 7.6×10^{11} N m^{-2}
3. 1.8×10^{10} N m^{-2}
4. 47 N, 7×10^{-3} J
5. 3.1 mm
6. 45 rev s^{-1}
7. 3870 N
10. 1.6×10^{11} N m^{-2}
11. 18 N
12. 354 J m^{-3}
13. 40 mm

EXERCISE 8 (p. 14)

1. 0.44
2. 0.3
4. 37 N
5. 1.7×10^{-4} N
6. ML^{-1} T^{-1}, N s m^{-2}
8. 16 : 27
9. 3.3 N s m^{-2}
11. 1.04×10^{-3} mm
13. 1.0×10^{-3} N s m^{-2}

ANSWERS

EXERCISE 9 (p. 15)

1. $N\,m^{-1}$, MT^{-2}, decreases
2. 7.7×10^{-3} N
3. 2.5×10^{-2} $N\,m^{-1}$
4. 7.5×10^{-2} $N\,m^{-1}$, 0.35 mm
5. 68 mm
6. 0.58 : 1
7. 0.133 m
8. 11 mm
9. (i) 5 $N\,m^{-2}$; (ii) 40 $N\,m^{-2}$
10. 4 mm
11. 12.1 cm
15. greater
16. minimum, sphere
17. (i) vaporization; (ii) critical
18. (ii) 50 mm
19. 8×10^{-2} J

EXERCISE 10 (p. 17) – Multiple Choice

1. B
2. E
3. C
4. D
5. D
6. D
7. A
8. C
9. D
10. D
11. C
12. A
13. B
14. E
15. A
16. C
17. (a) (i) D, (ii) A; (b) (i) C, (ii) E; (c) (i) E, (ii) B
18. D
19. (a) B; (b) D; (c) C; (d) E
20. (a) D; (b) A; (c) E; (d) A; (e) D; (f) B
21. B

HEAT

EXERCISE 11 (p. 26)

1. (i) 24 000 J; (ii) 5.6 K
2. 1.5 K
3. 0.24 K, 0.12 K
4. 60 p
5. 4100 $J\,kg^{-1}\,K^{-1}$
6. 4 K
7. 12.6 K
8. 12 600 J, 12.6 K
9. 1200 $J\,kg^{-1}\,K^{-1}$, less
10. (i) 60 W; (ii) 450 $J\,kg^{-1}\,K^{-1}$
11. 28°C
12. 2400 $J\,kg^{-1}\,K^{-1}$
13. 1100 $J\,kg^{-1}\,K^{-1}$
15. 7.9 g
16. 52°C
17. 0.028 kg
18. 17°C
19. 7 g
20. 8.8×10^5 $J\,kg^{-1}$, 2.7 W

EXERCISE 12 (p. 28)

1. (i) 153 cm^3; (ii) 137 cm^3; (iii) 103 cm^3
2. (i) 712.5; (ii) 682.5; (iii) 557.5 mmHg
3. 22.6 l
4. 1.7×10^5 $N\,m^{-2}$
5. (i) 0.37 g; (ii) 1.12 g
6. 88 mm
7. 0.92 m
8. 17.4°C
9. −272°C
10. 0.28 g
11. (i) 33; (ii) 80
12. 8.3 $J\,mol^{-1}\,K^{-1}$; 24 g, 12 mol
14. 1.0×10^{-4} kg
15. (i) 290 $J\,kg^{-1}\,K^{-1}$; (ii) 4100 $J\,kg^{-1}\,K^{-1}$; 1440 $J\,K^{-1}$
16. (i) 1, n, m/M mol
17. 8.3 $J\,mol^{-1}\,K^{-1}$
18. 4×10^{-5} kg

19. 0.24 mol
20. 4.75×10^4 N m^{-2}
21. 0.08 kg
22. 780 mmHg
23. 830 mmHg

EXERCISE 13 (p. 30)

1. (i) $N/3$; (ii) $N/6$
3. (i) 1840 m s^{-1}; (ii) 1930 m s^{-1}
5. 460 m s^{-1}, 540 m s^{-1}
6. (i) 2000 m s^{-1}; (ii) 1200 K; (iii) 262.5 K
8. 1304 m s^{-1}, 461 m s^{-1}
10. 2.9×10^{22}

EXERCISE 14 (p. 32)

1. 6000 J
3. (i) 12.45 J mol^{-1} K^{-1}; (ii) 622.5 J kg^{-1} K^{-1}
4. (i) 160 mols; (ii) 6×10^5 J
5. 20.9 J mol^{-1} K^{-1}
6. 8.4 J mol^{-1} K^{-1}, 0.09 kg m^{-3}
8. (i) 311 J kg^{-1} K^{-1}, 518 J kg^{-1} K^{-1}; (ii) 280 J K^{-1}
9. (i) 8.25 J mol^{-1} K^{-1}; (ii) 3100 J kg^{-1} K^{-1}; (iii) 9280 J
10. 7.4×10^6 J
12. (i) 169 J; (ii) 2091 J
13. 3000 J
14. (i) 3000 J; (ii) 12 000 J; (iii) 9000 J
15. 9.3×10^5 J K^{-1}, 3.5×10^5 J
17. 950, 1040 mmHg
18. (i) 1980 mmHg; (ii) 87°C
19. -6.5°C
21. (i) C_V, C_V, 0
    (ii) C_p, C_V, R
    (iii) 0, $-C_V \delta T$, $C_V \delta T$
    (iv) $p \delta V$, 0, $p \delta V$
    (v) ml, $ml - p \delta V$, $p \delta V$
    (vi) ml, ml, 0

EXERCISE 15 (p. 34)

2. 851 mmHg
3. 152 mmHg
6. 108 mmHg
8. 194 mmHg
9. 1.06 g
10. 166 cm^3

EXERCISE 16 (p. 36)

1. (i) 880 K m^{-1}; (ii) 1.32×10^6 J
2. 3.84×10^6 J
3. 390 W m^{-1} K^{-1}
4. 25°C
5. 410°C
6. 20 736 J, 6100 J
8. 91°C, 29°C
9. 1.4 W m^{-1} K^{-1}
11. 55 g

EXERCISE 17 (p. 37)

2. 3.3×10^9 J
3. 5763 K
4. 0.06 K s^{-1}
5. (i) 100 W; (ii) 2574 K
7. 0.5 K min^{-1}
8. 5790 K

EXERCISE 18 (p. 39)

1. (i) 30.0324 m; (ii) 29.9964 m; 7.2 mm
2. 2.00012 s
3. (i) 75 000 N; (ii) 7500 N
4. 46 s
5. 20 160 N
7. (i) 249.85 m^2; (ii) 250.22 m^2
8. (i) 13 360 kg m^{-3}; (ii) 13 610 kg m^{-3}
9. 5.5×10^{-4} K^{-1}
10. 35.6 cm
12. 1.9×10^{-7} m^3
13. 756 mm

ANSWERS

EXERCISE 19 (p. 40)

1. (i) 312 K; (ii) 219 K
2. 291 K, 237 K
3. 297 K
4. (i) 80°C, (ii) 120°C
5. (i) 59.6°C, (ii) −20.4°C
6. (i) 18°C; (ii) −45°C
10. 54°C

EXERCISE 20 (p. 41) – Multiple choice

1. B
2. A
3. C
4. E
5. D
6. A
7. A
8. C
9. D
10. B
11. E
12. C
13. A
14. D
15. (i) C; (ii) D; (iii) C; (iv) A; (v) B; (vi) E; (vii) D
16. (a) B; (b) C; (c) C; (d) E; (e) A
17. (a) B; (b) E; (c) A
18. (a) D; (b) B
19. (a) C; (b) A; (c) E; (d) D; (e) B; (f) A; (g) C
20. C
21. (a) E; (b) A; (c) C; (d) B

GEOMETRICAL OPTICS

EXERCISE 21 (p. 51)

2. 1 m
5. 6
8. (i) 30 cm, $m = 5$; (ii) 7.5 cm, $m = 2.5$
9. 0.15 m
10. (i) 10 cm, $m = 4/9$; (ii) 6 cm, $m = 2/3$
11. 16 cm
12. 1 m
13. $3b$
14. 0.4 m
15. 30 cm, 2.5 cm

EXERCISE 22 (p. 52)

2. (i) 22.5°; (ii) 59.4°; (iii) 30.6°
4. 3.3 cm, 4.8 cm
7. (i) 41.5°; (ii) 48.8°; (iii) 61.8°
10. 4.86 cm
11. 0.3 cm

EXERCISE 23 (p. 53)

1. 52.4°
2. 1.56
3. 38.9°
5. 27.9°
9. 61.5°
10. 29.4°, 59.4°
11. 82.3°
12. 75.5°

EXERCISE 24 (p. 54)

1. (i) 60 cm; (ii) 12 cm
2. 26.7 cm, 13.3 cm
3. (i) 6.7 cm, (ii) 3.3 cm
4. 90 cm
5. 2.3 cm
6. (i) 20 cm; (ii) 16.7 cm
7. (i) 7.2 cm; (ii) 36 cm
9. 3.5 cm
10. 4 mm
13. 24 cm
14. 10 cm
17. 24 cm
18. 48 cm
19. 20.4 cm, 42.9 cm
20. 1.38
21. 450 mm
22. −33.3 cm

23. $f = +27.8$ cm, 25 to 38.5 cm
24. $f = -150$ cm, 23.1 cm
25. $f = -24$ cm, 48 cm to ∞
26. 66.7 cm, 13.6 cm
27. $f = -300$ cm, 15.8 cm

EXERCISE 25 (p. 56)

1. $3.18°, 0.12°$
2. $2.56°, 0.16°$
3. 0.038, 0.063
4. $4°, 1.68°$
5. $6.04°, 0.08°$
8. 23.1 cm, 22.2 cm
9. 48.5 cm, 51.6 cm
11. -132.5 cm, 202 cm
13. 1.04 cm, 0.96 cm
14. $+79.2$ cm (crown), -131.3 cm (flint)

EXERCISE 26 (p. 58)

1. $3 : 1$
2. (i) 6; (ii) 7.2
3. 2.5 cm, 102.5 cm, 44
4. (i) 5 cm; (ii) 4 mm
7. 3.5 cm
8. (i) 6.25; (ii) 7.25
9. 9.4 cm, 14.5
10. 19.2 cm, 30
11. 4.7 cm, 32.8
12. 60.8 cm, 43.3
13. 6.4; 4.4 cm
14. 12 cm
15. 0.3 m
16. 2.4 mm; $1.5 : 1$
17. $1/256$ s

PHYSICAL OPTICS

EXERCISE 27 (p. 60)

1. (i) 3.0×10^{-7} m, (ii) 4.5×10^{-7} m
4. 2.82×10^6 m s^{-1}
5. 1.97×10^8 m s^{-1}, 2.00×10^8 m s^{-1}; blue

9. 4.5×10^{-7} m
10. 16 cm
13. (i) 5.5×10^{-5} s; (ii) $1/1440$ rev; (iii) 3.13×10^8 m s^{-1}
14. 3×10^5 km s^{-1}
16. $300\,800$ m s^{-1}
17. 3×10^8 m s^{-1}

EXERCISE 28 (p. 62)

1. 5.4×10^{-7} m
2. 0.45 mm, 0.75 mm
6. 0.019 mm
7. (i) 1.5×10^{-3} rad; (ii) 0.16 mm
8. 5.89×10^{-7} m
11. 2.65 m
12. (i) 20 bands shift; (ii) 0.23 mm apart; (iii) 0.25 mm apart
14. 5.44×10^{-7} m

EXERCISE 29 (p. 63)

1. $17° 28'$; 3
2. $1° 33'$; 13.5 mm
3. 5.72×10^{-7} m; $20° 4'$
6. 4th and 3rd orders; $46° 3'$
9. (i) 7.3×10^{-7} rad; (ii) 1.4×10^{-7} rad
10. (ii) 3.2×10^{-3} rad
11. (i) 1; (ii) $4 : 1$; (iii) $1 : 4$

EXERCISE 30 (p. 64)

3. $57°$

EXERCISE 31 (p. 65) – Multiple Choice

1. D
2. C
3. A
4. B
5. E
6. C
7. A
8. B

ANSWERS

9. C
10. D
11. D
12. (a) A; (b) E; (c) B; (d) D; (e) A; (f) C
13. B
14. (a) D; (b) C; (c) B; (d) A; (e) E; (f) B; (g) E
15. C
16. D
17. A
18. E
19. D
20. A
21. B
22. C
23. B
24. D
25. D
26. A
27. B
28. E
29. D
30. A
31. C
32. B
33. E
34. D

EXERCISE 32 (p. 75)

1. (i) $2\pi/3$; (ii) $6\pi/5$; (iii) $2\pi/3$ rad
2. 20 Hz, 0.63 m s^{-1}
3. (i) $3\pi/5$ rad; (ii) 0
5. $T = \lambda/v$; $y = a \sin 2\pi (vt + x)/\lambda$
7. ω/β, $2\pi/\beta$, $2\pi/\omega$
8. 1000 Hz, 0.34 m, 340 m s^{-1}
9. $y = 2 \times 10^{-3} \sin 4\pi (10^4 t - x)$, y and x in m
10. (i) B; (ii) C; (iii) F; (iv) D; (v) A
11. 500 m s^{-1}
12. 0.035 s
13. 550 Hz
14. 320 m s^{-1}

17. (i) 100 Hz; (ii) 3.1 m; (iii) 10^{-6} m; (iv) 1.6 m; (v) (a) 0; (b) π rad

EXERCISE 33 (p. 77)

6. nodes, antinodes, NA = 0.17 m
7. (i) f same, zero phase diff.; (ii) f 1 : 2; (iii) f same, phase diff. 90°
8. 6.3 W, 1.25×10^{-3} W m^{-2}
9. 0.2 mm, 4×10^{-5} W m^{-2}
10. (i) 850 Hz; (ii) 340 Hz
11. (i) 341 m s^{-1}; (ii) 332 m s^{-1}; 344 m s^{-1}
12. 1.13 m
13. 331 m s^{-1}
15. 66 m
16. 4 s^{-1}
17. 509 Hz, 515 Hz
19. 877 Hz, 735 Hz
20. 958 Hz, 1097 Hz
21. (i) 979 Hz; (ii) 938 Hz
22. 5.3 Hz, 5.3 Hz
23. 3.2 Hz
24. 24 m s^{-1}, 26 m s^{-1}

EXERCISE 34 (p. 79)

1. 140 Hz, 420 Hz
2. 340 Hz, 680 Hz
3. 316 m s^{-1}, 6 mm
4. 171 Hz
7. 337 m s^{-1}; 168 Hz, 165 Hz
9. (i) 200 m s^{-1}; (ii) 250 Hz; (iii) 750 Hz
10. 16 : 9
11. 5 s^{-1}
12. 0.33 mm, 300 Hz
14. 20 N
15. 50 Hz
16. (i) 1000 m s^{-1}; (ii) 6000 m s^{-1}
17. 6844 m s^{-1}, 4×10^{11} N m^{-1}
18. 11 333 m s^{-1}; 453 m s^{-1}
20. 1700 Hz. (i) $4\frac{1}{2}\lambda$; (ii) 1889 Hz

EXERCISE 35 (p. 81) – Multiple Choice

1. B
2. A
3. E
4. B
5. C
6. D
7. D
8. (a) B; (b) C; (c) D; (d) B; (e) D; (f) A

ELECTROSTATICS

EXERCISE 36 (p. 86)

1. 9×10^{-5} N; 6×10^{-2} m
2. (i) 1.8×10^{8} V m^{-1}; (ii) 1.44×10^{-10} N
7. (i) 1.6×10^{-10} C
8. (i) 1.4×10^{11} V m^{-1}; (ii) -1.8×10^{8} V
9. 900 V; A to B
10. (i) 1000 V; (ii) 2×10^{5} V m^{-1}; (iii) 2×10^{-4} N
11. (i) 5×10^{4} V m^{-1}; (ii) 1.9×10^{7} m s^{-1}
12. 8×10^{-10} J, 4.2×10^{7} m s^{-1}
13. 7.3×10^{6} m s^{-1}
15. 8×10^{-7} C, 3.6×10^{12} Ω
16. 7.7×10^{-15} kg
17. (i) 5×10^{4} V m^{-1}; (ii) 2.4×10^{-17} J; (iii) 7.3×10^{6} m s^{-1}
20. (i) 1.9×10^{-7} C m^{-2}; (ii) 0; (iii) 1080 V; (iv) 21 600 V m^{-1}
21. A = 3600 V; B = 900 V; C = 1800 V

EXERCISE 37 (p. 88)

1. 1×10^{-7} F, 4×10^{-6} C
2. 40 V; 13.3 V, 2.7×10^{-5} C, 5.3×10^{-5} C
3. 4×10^{-8} F
4. 8×10^{-7} A
5. 600 V, 1.2×10^{-6} J
6. 9×10^{-12} F, 3×10^{-7} C
7. 6.8×10^{-8} C, 1.0×10^{-5} J; 3.0×10^{-5} J
8. 9×10^{-12} F m^{-1}
9. (i) 1.2 μF; (ii) 2.2 μF
10. 1800 V; 360 V, 8×10^{-10} C
11. 10^{-2} J; 6.3×10^{-4} J, 1.9×10^{-3} J
12. (i) 4.8×10^{-4} C, 7.2×10^{-4} C; (ii) 2.9×10^{-4} C, 2.9×10^{-4} C; 1.7×10^{-2} J
13. 1.2×10^{-4} C each, 4.8 V, 1.2 V: (i) 1.92 V, (ii) 1.3×10^{-4} J
14. (i) 4.4×10^{-8} J; (ii) 8.8×10^{-8} J
15. (i) 10^{-7} C, 10^{-5} J; (ii) 4×10^{-10} F, 1.25×10^{-5} J
16. (i) 5×10^{-5} A; (ii) 4×10^{-4} C; (iii) 16 s
17. (i) 2.5×10^{-4} C; (ii) 5×10^{-6} A; (iii) 2×10^{-6} A
21. 1.1×10^{-10} F, 89 V
22. 4000 V m^{-1}, 1.5×10^{5} V
23. (i) 140 V; (ii) 7×10^{-8} J
24. 1.2×10^{-4} J. (i) 4500 V; (ii) 2.5×10^{-8} C, 2×10^{-8} C; (iii) 1.0×10^{-4} J
25. 5.6×10^{-8} C, 2.8×10^{-6} J
26. (i) 1.3×10^{-9} F; (ii) 2.7×10^{-5} J

EXERCISE 38 (p. 91) – Multiple Choice

1. C
2. D
3. B
4. A
5. C
6. E
7. D
8. A
9. B
10. D
11. (a) A; (b) E; (c) D
12. (a) A; (b) C

ANSWERS

CURRENT ELECTRICITY

EXERCISE 39 (p. 95)

1. (i) 0.15 V; (ii) 0.2 Ω; (iii) 0.2 A
2. 200 V
3. (i) 1.2, 0.3, 0.9 A; (ii) 102, 18, 18 V
4. 4, 3, 2 A
5. (i) 0.6 A; (ii) 9.6 V
6. (i) 0.5 A; (ii) 2 Ω
7. (i) 1 A, 4 V; (ii) 1.2 A, 2.4 V
8. (i) 0.5 A; (ii) 3 V, 4.5 V
9. 0.1 A
10. (i) 1 A; (ii) 0.5 A; (iii) 8 V
11. (i) 0.3 A; (ii) 3.7 V, 1.1 V; (iii) +3 V, −1.8 V
13. (i) C; (ii) D; (iii) A; (iv) B
16. 1.6×10^{-4} m s^{-1}
17. 6 Ω, 1/3 A
18. 12 V, 1 A

EXERCISE 40 (p. 97)

1. 960 Ω; 403 p; 726 MJ
2. 6.4 W, 1.6 W, 80%
3. 3.5 A; 480 Ω, 576 Ω; 508 MJ
4. 18°C
5. 2200 J kg^{-1} K^{-1}
7. 10.4 mm
8. 60 A
9. 1 kW
10. 8 Ω

EXERCISE 41 (p. 98)

1. 6.0 Ω
3. 6.7 Ω
4. 72.7 cm
6. (i) 9 Ω; (ii) 0.13 A; (iii) 1.2 V
7. 8×10^{-8} Ω m
8. 0.32 m
9. 43.2 Ω
10. 0.63 mm
11. 3.9×10^{-3} K^{-1}, 177°C
12. 3.3×10^{-3} K^{-1}, 3.11 Ω
15. (i) 0.101 Ω, (ii) 4%

EXERCISE 42 (p. 99)

1. 1.47 V
2. 0.16 A
5. (i) 62.5 cm; (ii) 1 Ω
7. 8.8 Ω
8. 9995 Ω

EXERCISE 43 (p. 101)

1. 3.3×10^{-7}, 1.1×10^{-6} kg C^{-1}
2. 0.05 A
3. 3.3×10^{-7} kg C^{-1}, 0.031 g
4. 0.49 A
5. 0.021, 2.27, 0.66 g
6. 96 500 C 193 000 C
7. 3.0 A
8. 1.7×10^{-27} kg
9. 1.56 V
11. 32.6 Ω, 250 W, 1.25 W

EXERCISE 44 (p. 102)

1. (i) flux; (ii) flux density; (iii) 1 Wb m^{-2}
2. 8×10^{-3} N
3. (i) 0.01 N; (ii) 0.005 N; (iii) 0
4. (i) 6.4×10^{-4} N m; (ii) 3.2×10^{-4} N m
5. (i) 4×10^{-5} T; (ii) 4×10^{-4} N m^{-1}
7. (i) shunt 3.34×10^{-2} Ω; (ii) series 1990 Ω; (iii) shunt 1.67×10^{-2} Ω; (iv) series 390 Ω
9. 5×10^{-7} N m rad^{-1}
10. 2.4×10^{-3} N
11. (i) series 24 500 Ω; (ii) shunt 55.6 Ω
13. (i) 2.4°; (ii) 0.08°; (iii) 2° mA^{-1}
17. (i) Y; (ii) X
20. (i) Bev; (ii) V_H/b; (iii) eV_H/b; $V_H = Bvb$
21. 5×10^{-7} V (approx)

EXERCISE 45 (p. 104)

1. (i) 0.16 V; (ii) 0.2 V
2. 0.16 V
6. (i) 4×10^{-5} V; (ii) 0
7. 9.9×10^{-4} V
8. 6.3×10^{-2} V, 0
9. (i) 251 V; (ii) 218 V; (iii) 0
11. (i) 120 μC; (ii) 240 μC
15. 6.4×10^{-4} T
16. 1.5 H; 2000 turns
17. (i) 10^{-3} T; (ii) 6×10^{-5} H
18. 0.02 H, 0.04 Wb
19. (i) 4 A s^{-1}; (ii) 3 A s^{-1}

EXERCISE 46 (p. 106)

1. (i) narrow circular; (ii) straight; (iii) solenoid
3. 8×10^{-6} T; 2×10^{-5} T
6. 5×10^{-4} T, 1.5×10^{-5} T
8. (i) 1.6×10^{-5} T; (ii) 3.2×10^{-6} N

EXERCISE 47 (p. 107)

6. 0.3 T, 478

EXERCISE 48 (p. 108)

1. (i) 5 A; (ii) 3.5 A; (iii) 2.5 A; (iv) 50 W
2. (i) 8 A; (ii) 5.7 A; (iii) 1.1 A; (iv) 128 W
3. (i) 2 V; (ii) 6.4 V
4. (i) 1257 Ω; (ii) 796 Ω
5. 2 A, 0.32 A
6. 0.3 V, 0.4 V
7. (i) 0.02 A; (ii) 6 V, 8 V; (iii) 37°; (iv) 0.16 W
8. (i) 9.6×10^{-3} A; (ii) 9.6 V, 2.9 V; (iii) 73°; (iv) 0.028 W
9. (i) 0.04 A; (ii) 4000 Ω; (iii) 10 H; (iv) 0.16 W
10. 12.5 H, 2.5×10^{-7} F
11. 0.11 A, 5 W
14. 1.4 V
15. (i) 8×10^4 Hz; (ii) 4×10^{-3} A; (iii) 40 V; (iv) 8×10^{-4} W
17. (i) 5 V; (ii) 1000 V, 1000 V
18. 6.3×10^{-2}. 3.2×10^{-2} A; 3.1×10^{-2} A

ATOMIC PHYSICS

EXERCISE 49 (p. 114)

1. 1/1848
2. 4.8×10^{-15} N, 5.4×10^{15} m s^{-2}
3. 4000 V
4. 0.14 m
5. 5×10^{-3} T
6. 7.3×10^6 m s^{-1}, 0.41 m
7. 1.9×10^{-14} kg, 1.7×10^{-3} mm
8. 24.8 mm
9. 1.25×10^{-4} m s^{-1}
10. 10^7 m s^{-1}, 5.8×10^{-2} m
11. (i) 2×10^{-3} mm; (ii) 4.8×10^{-19} C
12. 8×10^6 m s^{-1}, 1.7×10^{11} C kg^{-1}
13. 1.9×10^{11} kg C^{-1}
14. 5×10^{16}. (i) 1.9×10^7 m s^{-1}; (ii) 8.6×10^{-7} N
15. $v = Bqr/M$; 3.1×10^{-8} s, 3.2×10^7 m s^{-1}
18. (i) 1.6×10^{-4} A; (ii) 20 T

EXERCISE 50 (p. 116)

5. (i) 6250 Ω; (ii) 1.5
8. (i) 15 km; (ii) 400 Hz
9. 50 Hz
20. (i) p-n-p; (ii) isolates input d.c. from base; (iii) YZ; (iv) C_3; (v) (a) 3.3 k, (b) 1 k, (c) 220 k
21. (i) n-p-n; (ii) isolates input d.c. from base; (iii) MN; (iv) R_4; (v) (a) R_1, (b) R_4, (c) R_3

EXERCISE 51 (p. 119)

1. (i) 2.6×10^{-19} J; (ii) 4.6×10^{-19} J;

ANSWERS

(iii) 6.6×10^{-19} J
2. (i) 3.3×10^{-19} J;
(ii) 4.4×10^{-19} J
3. 2.8×10^{-19} J
4. 1.1×10^{-19} J
5. 6.9×10^{-7} m, 1.5×10^{-19} J, 0.94 V
6. 5.05×10^{-7} m, 0.46×10^{-19} J, 0.29 V
11. (i) 6.3×10^5 m s^{-1}; (ii) 1.1 V; (iii) 6.7×10^{-7} m
13. 1.2×10^8 m s^{-1}, 6.4×10^{-15} J
14. (i) 2.0×10^{-15} J; (ii) 12 400 eV
15. 0.2×10^{-10} m, 0.4×10^{-10} m
18. 0.6×10^{-10} m
19. (i) 14.5°; (ii) 38.7°
22. 0.4×10^{-10} m
23. 31 000 V, 2.9×10^{-10} m

EXERCISE 52 (p. 121)

1. 2.5×10^{-7} m
4. (i) 6.73×10^{-7} m, 1.23×10^{-7} m;
(ii) 13.4 V
5. $-0.38, -0.55, -0.85, -1.51$ eV

EXERCISE 53 (p. 122)

3. 60.7 : 1.4 : 1
6. 168 s
7. 75 s (approx)
10. (i) 56 days; (ii) 80 days
11. (i) 95%; (ii) 3.6×10^7 d.p.s.
12. 8×10^9 s^{-1}, 173 s
14. (i) 207, 82; (ii) n to p, p to n
15. He: 2, 2, 2; N: 7, 7, 7, U: 92, 92, 146
16. oxygen
17. 9×10^{10} J
18. 1.5×10^{-10} J, 937.5 MeV
19. 0.0076 u
20. 0.022 u
21. 0.18 u
22. (i) 13 : 24; (ii) 0.195 m
25. (i) 23.8 MeV; (ii) 28.3 MeV

EXERCISE 54 (p. 129)

1. D
2. A
3. D
4. E
5. C
6. B
7. C
8. D
9. E
10. C
11. B
12. D
13. (a) C; (b) D; (c) B; (d) A
14. C
15. D
16. A
17. D
18. (a) B; (b) E; (c) D
19. (a) C; (b) E; (c) A; (d) D
20. B
21. D
22. A
23. C
24. C
25. C
26. A
27. B
28. D
29. E
30. B
31. C
32. D
33. A
34. C
35. E
36. D
37. C
38. C
39. A
40. D
41. C
42. E
43. B

44. A
45. C
46. E
47. B
48. C
49. E
50. A
51. D
52. C
53. C
54. E

EXERCISE 55 (p. 135)

1. B
2. E
3. A
4. D
5. C
6. B
7. J
8. A
9. C
10. K
11. N
12. J
13. L/H
14. H
15. A
16. G
17. A
18. J
19. E
20. O
21. E
22. D
23. M
24. E
25. F
26. (*a*) B; (*b*) A; (*c*) E; (*d*) B; (*e*) C; (*f*) A
27. (*a*) C; (*b*) E; (*c*) A; (*d*) D; (*e*) A
28. B
29. D
30. F
31. A
32. E
33. C
34. B

EXERCISE 56 (p. 138)

1. C
2. B
3. D
4. A
5. E
6. B
7. C
8. D
9. B
10. A
11. E
12. A
13. D
14. B
15. A
16. D
17. C
18. B
19. B
20. A
21. D
22. B
23. E
24. B
25. E
26. A
27. D
28. C
29. C
30. B